Envelopes and Sharp Embeddings of Function Spaces

CHAPMAN & HALL/CRC
Research Notes in Mathematics Series

Submission of proposals for consideration

Suggestions for publication, in the form of outlines and representative samples, are invited by the Editorial Board for assessment. Intending authors should approach one of the main editors or another member of the Editorial Board, citing the relevant AMS subject classifications. Alternatively, outlines may be sent directly to the publisher's offices. Refereeing is by members of the board and other mathematical authorities in the topic concerned, throughout the world.

Preparation of accepted manuscripts

On acceptance of a proposal, the publisher will supply full instructions for the preparation of manuscripts in a form suitable for direct photo-lithographic reproduction. Specially printed grid sheets can be provided. Word processor output, subject to the publisher's approval, is also acceptable.

Illustrations should be prepared by the authors, ready for direct reproduction without further improvement. The use of hand-drawn symbols should be avoided wherever possible, in order to obtain maximum clarity of the text.

The publisher will be pleased to give guidance necessary during the preparation of a typescript and will be happy to answer any queries.

Important note

In order to avoid later retyping, intending authors are strongly urged not to begin final preparation of a typescript before receiving the publisher's guidelines. In this way we hope to preserve the uniform appearance of the series.

CRC Press, Taylor and Francis Group
24-25 Blades Court
Deodar Road
London SW15 2NU
UK

Dorothee D. Haroske

Envelopes and Sharp Embeddings of Function Spaces

CRC Press
Taylor & Francis Group
Boca Raton London New York

CRC Press is an imprint of the
Taylor & Francis Group, an **informa** business

A CHAPMAN & HALL BOOK

CRC Press
Taylor & Francis Group
6000 Broken Sound Parkway NW, Suite 300
Boca Raton, FL 33487-2742

First issued in paperback 2019

© 2007 by Taylor & Francis Group, LLC
CRC Press is an imprint of Taylor & Francis Group, an Informa business

No claim to original U.S. Government works

ISBN-13: 978-1-58488-750-8 (hbk)
ISBN-13: 978-0-367-39031-0 (pbk)

**Visit the Taylor & Francis Web site at
http://www.taylorandfrancis.com**

**and the CRC Press Web site at
http://www.crcpress.com**

To my family

Contents

Preface

We present the new concept of growth envelopes and continuity envelopes in function spaces. This originates from such classical results as the famous Sobolev embedding theorem [Sob38], or, secondly, from the Brézis-Wainger result [BW80] on the almost Lipschitz continuity of functions from a Sobolev space $H_p^{1+n/p}(\mathbb{R}^n)$, $1 < p < \infty$. In the first case questions of growth are studied: what can be said about the unboundedness of functions belonging to, say, $H_p^{n/p}(\mathbb{R}^n)$, $1 < p < \infty$? We introduce the *growth envelope* $\mathfrak{E}_{\mathsf{G}}(X) = (\mathcal{E}_{\mathsf{G}}^X(t), u_{\mathsf{G}}^X)$ of a function space $X \subset L_1^{\mathrm{loc}}$, where

$$\mathcal{E}_{\mathsf{G}}^X(t) \sim \sup \{ f^*(t) : \|f|X\| \le 1 \}, \quad 0 < t < 1,$$

is the *growth envelope function* of X and $u_{\mathsf{G}}^X \in (0, \infty]$ is some additional index providing an even finer description of the unboundedness of functions belonging to X.

Instead of investigating the growth of functions one can also focus on their smoothness, i.e., for $X \hookrightarrow C$ it makes sense to replace $f^*(t)$ with $\frac{\omega(f,t)}{t}$, where $\omega(f,t)$ is the modulus of continuity. The *continuity envelope function* $\mathcal{E}_{\mathsf{C}}^X$ and the continuity envelope $\mathfrak{E}_{\mathsf{C}}$ are introduced completely parallel to $\mathcal{E}_{\mathsf{G}}^X$ and $\mathfrak{E}_{\mathsf{G}}$, respectively, and similar questions are studied.

These concepts are first explained in detail and demonstrated on some classical and rather obvious examples in Part I; in Part II we deal with these instruments in the context of spaces of Besov and Triebel-Lizorkin type, $B_{p,q}^s$ and $F_{p,q}^s$, respectively.

In the end we turn to some applications, e.g., concerning the asymptotic behaviour of approximation numbers of (corresponding) compact embeddings. Further applications are connected with Hardy-type inequalities and limiting embeddings. We discuss the relation between growth and continuity envelopes of a suitable pair of spaces. Problems of global nature are regarded, and we study situations where the envelope function itself belongs to or can be realised in X, respectively.

I am especially grateful to Professor Hans Triebel; while he was preparing his book [Tri01] (in which Chapter 2 is devoted to related questions) we could discuss the material at various occasions. This led to the preprint version [Har01], and also became part of [Har02]. But for some reason these results –

though extended, improved, used, cited already – were never published else-where. In view of the substantial material, the idea appeared later to collect it in a book rather than a number of papers. This gives me the opportunity for special thanks to Professor David E. Edmunds who offered invaluable mathematical and linguistic comments, and to Professor Haïm Brézis who encouraged me in that form of publication. Last but not least I appreciate joint work and exchange of ideas on the subject with many colleagues, in particular Professor António M. Caetano and Dr. Susana D. Moura. Finally, I am indebted to my family for their never-ending patience and support.

Jena, March 2006   *Dorothee D. Haroske*

Part I

Definition, basic properties, and first examples

Chapter 1

Introduction

We present our recently developed concept of envelopes in function spaces – a relatively simple tool for the study of classical, and also rather complicated spaces, say, of Besov or Triebel-Lizorkin type, respectively, in so-called "*limiting*" situations. This subject is still very new, but in our opinion it has grown to such a degree of maturity that it is now worth a coherent account. The topic is studied in two steps: on a more general level, in Part I, where we do not assume any knowledge of the scales of function spaces mentioned above, and subsequently, in Part II, the results are essentially related to spaces of type $B_{p,q}^s$ and $F_{p,q}^s$. This also explains the main structure of this book.

We first describe the common background for both parts and explicate the programme afterwards. In fact, considerable parts of the outcomes were obtained much earlier and already summarised in the long preprint [Har01]; but for some reason they have not yet been published (apart from [Har02], essentially relying on [Har01]). However, we also complemented and extended the presentation [Har01] quite recently.

The history of such questions starts in the 1930s with Sobolev's famous embedding theorem [Sob38]

$$W_p^k(\Omega) \hookrightarrow L_r(\Omega), \tag{1.1}$$

where $\Omega \subset \mathbb{R}^n$ is a bounded domain with sufficiently smooth boundary, L_r, $1 \leq r \leq \infty$, stands for the usual Lebesgue space, and W_p^k, $k \in \mathbb{N}$, $1 \leq p < \infty$, are the classical Sobolev spaces. The latter spaces have been widely accepted as one of the crucial instruments in functional analysis – in particular, in connection with PDEs – and have played a significant role in numerous parts of mathematics for many years. Sobolev's famous result (1.1) holds for $k \in \mathbb{N}$ with $k < \frac{n}{p}$, and r such that $\frac{k}{n} - \frac{1}{p} \geq -\frac{1}{r}$ (strictly speaking, [Sob38] covers the case $\frac{k}{n} - \frac{1}{p} > -\frac{1}{r}$, whereas the extension to $\frac{k}{n} - \frac{1}{p} = -\frac{1}{r}$ was achieved later). In the limiting case, when $k = \frac{n}{p} \in \mathbb{N}$, the inclusion (1.1) does not hold for $r = \infty$, whereas for all $r < \infty$,

$$W_p^{n/p}(\Omega) \hookrightarrow L_r(\Omega). \tag{1.2}$$

The theory of Sobolev-type embeddings originates in classical inequalities from which integrability properties of a real function can be deduced from those of its derivatives. In that sense (1.2) can be understood simply as the impossibility of specifying integrability conditions of a function $f \in W_p^{n/p}(\Omega)$ merely by means of L_r conditions. In order to obtain further refinements of the limiting case of (1.1) it becomes necessary to deal with a wider class of function spaces. In the late 1960s Peetre [Pee66], Trudinger [Tru67], Moser [Mos71], and Pohožaev [Poh65] independently found refinements of (1.1) expressed in terms of Orlicz spaces of exponential type, see also Strichartz [Str72]; this was followed by many contributions in the last decades investigating problems related to (1.1) in detail. In 1979 Hansson [Han79] and Brézis, Wainger [BW80] showed independently that

$$W_p^{n/p}(\Omega) \hookrightarrow L_{\infty,p} (\log L)_{-1} (\Omega), \tag{1.3}$$

where $1 < p < \infty$, and the spaces $L_{r,u} (\log L)_a (\Omega)$ appearing in (1.3) provide an even finer tuning, see also Hedberg [Hed72], and sharper results by Maz'ya [Maz72] and [Maz05]. Recently we noticed a revival of interest in limiting embeddings of Sobolev spaces indicated by a considerable number of publications devoted to this subject; let us only mention a series of papers by Edmunds with different co-workers ([EGO96], [EGO97], [EGO00], [EK95], [EKP00]), by Cwikel, Pustylnik [CP98], and – also from the standpoint of applications to spectral theory – the publications [ET95] and [Tri93] by Edmunds and Triebel. This list is by no means complete, but reflects the increased interest in related questions in the last years. There are a lot of different approaches to the modification of (1.1) in order to get – in the adapted framework – appropriately optimal assertions. Apart from the original papers, assertions of this type are indispensable parts in books dealing with Sobolev spaces and related questions, cf. [AF03], [Zie89], [Maz85], [EE87], [EE04], but there is a vast literature devoted to (extensions of) this subject.

Returning to the limiting case $k = \frac{n}{p}$ (1.2), for instance, one can also extend the scale of admitted (source) spaces in another direction, first replacing classical Sobolev spaces $W_p^{n/p}$ by the more general fractional Sobolev spaces $H_p^{n/p}$, or even by spaces of type $B_{p,q}^s$ and $F_{p,q}^s$, respectively. It is well-known, for instance, that $B_{p,q}^{n/p} \hookrightarrow L_\infty$ if, and only if, $0 < p < \infty$, $0 < q \le 1$ – but what can be said about the growth of functions $f \in B_{p,q}^{n/p}$ otherwise, i.e., when $B_{p,q}^{n/p}$ contains essentially unbounded functions ? Edmunds and Triebel proved in [ET99b] that one can characterise such spaces by sharp inequalities involving the non-increasing rearrangement f^* of a function f: Let $\varkappa$ be a bounded, continuous, decreasing function on $(0,1]$ and $1 < p < \infty$. Then there is a constant $c > 0$ such that

$$\left(\int_0^1 \left(\frac{f^*(t)\varkappa(t)}{|\log t|} \right)^p \frac{dt}{t} \right)^{1/p} \le c \, \left\| f | H_p^{n/p} \right\| \tag{1.4}$$

for all $f \in H_p^{n/p}$ if, and only if, $\varkappa$ is bounded; there are further results related to the case of H_p^s with $0 < s < \frac{n}{p}$ in [Tri99].

Recall that the scale $F_{p,q}^s$ of Triebel-Lizorkin spaces extends the (classical) Sobolev scale further, whereas Besov spaces $B_{p,q}^s$ have been well-known for a long time, either when characterised by differences or – nowadays preferably – in Fourier-analytical terms or via (sub-)atomic decompositions. They appear naturally in signal analysis, in contemporary harmonic analysis, in stochastics, and while studying approximation problems or solving PDEs; thus it is of deep interest to understand these spaces very well – apart from the purely functional analytic purpose. The theory of the scales $B_{p,q}^s$ and $F_{p,q}^s$ has been systematically studied and developed by Triebel in the last decades; his series of monographs [Tri78a], [Tri83], [Tri92], [Tri97], [Tri01], and the forthcoming book [Tri06] can be regarded as the most complete treatment of related questions.

Assertions of type (1.4) are already linked to our concept of envelopes in some sense. We realised that many contributions to the subject of limiting embeddings and sharp inequalities as (partly) mentioned above, share a little disadvantage – beside their unquestioned beauty: as far as characterisations of spaces of type $B_{p,q}^s$ or $F_{p,q}^s$ are concerned, they are usually bound to a certain setting. That is, dealing with such embeddings, one asks, say, for optimality of original or target spaces within a prescribed (fixed) context. We prefer to look for some feature "belonging" to the spaces under consideration only, and, moreover, defined as simply as possible (using classical approaches). This would enable us to gain significantly from the rich history and many forerunners. In view of the above-mentioned papers it was natural to choose the non-increasing rearrangement f^* as the basic concept on which our new tool should be built. This led us to the introduction of the *growth envelope function* of a function space X,

$$\mathcal{E}_{\mathsf{G}}^X(t) := \sup_{\|f|X\|\leq 1} f^*(t), \qquad 0 < t < 1. \tag{1.5}$$

It turns out that in rearrangement-invariant spaces there is a connection between $\mathcal{E}_{\mathsf{G}}^X$ and the fundamental function φ_X; we shall derive further properties and give some examples. The pair $\mathfrak{E}_{\mathsf{G}}(X) = \left(\mathcal{E}_{\mathsf{G}}^X(t), u_{\mathsf{G}}^X\right)$ is called the *growth envelope* of X, where u_{G}^X, $0 < u_{\mathsf{G}}^X \leq \infty$, is the infimum of all numbers v satisfying

$$\left(\int_0^\varepsilon \left[\frac{f^*(t)}{\mathcal{E}_{\mathsf{G}}^X(t)}\right]^v \mu_{\mathsf{G}}(\mathrm{d}t)\right)^{1/v} \leq c\,\|f|X\| \tag{1.6}$$

for some $c > 0$ and all $f \in X$, and μ_{G} is the Borel measure associated with

$-\log \mathcal{E}_{\mathsf{G}}^X$. The result reads for the Lorentz spaces $L_{p,q}$ as

$$\mathcal{E}_{\mathsf{G}}\left(L_{p,q}\right) = \left(t^{-\frac{1}{p}}, q\right),$$

and for Sobolev spaces W_p^k,

$$\mathcal{E}_{\mathsf{G}}\left(W_p^k\right) = \left(t^{-\frac{1}{p}+\frac{k}{n}}, p\right), \quad 1 \le p < \infty, \quad k < \frac{n}{p}. \tag{1.7}$$

We also deal with some weighted L_p-spaces, illuminating the difference between locally regular weights like $(1+|x|^2)^{\alpha/2}$, $\alpha > 0$, and corresponding Muckenhoupt A_p weights like $|x|^\alpha$, $0 < \alpha < n(1-\frac{1}{p})$. In Part II we consider characterisations for spaces of type $B_{p,q}^s$ and $F_{p,q}^s$, where $n(\frac{1}{p}-1)_+ \le s \le \frac{n}{p}$. Returning to our example (1.4) above one proves that

$$\mathcal{E}_{\mathsf{G}}\left(H_p^{n/p}\right) = \left(|\log t|^{1-\frac{1}{p}}, p\right), \quad 1 < p < \infty, \tag{1.8}$$

even in a more general setting. The counterpart for Besov spaces is given by

$$\mathcal{E}_{\mathsf{G}}(B_{p,q}^{n/p}) = \left(|\log t|^{1-\frac{1}{q}}, q\right), \quad 1 < q \le \infty, \quad 0 < p < \infty. \tag{1.9}$$

Unlike [Tri01, Ch. 2] where similar questions have been considered, we also study (some) borderline and weighted cases.

In an appropriately modified context it also makes sense to consider embeddings like (1.1) and (1.2) in "super-critical" situations, that is, when $k > \frac{n}{p}$. Then by simple monotonicity arguments all distributions are essentially bounded; moreover, one even knows that

$$W_p^k \hookrightarrow C \tag{1.10}$$

in this case, where C stands for the space of all complex-valued bounded uniformly continuous functions on $\mathbb{R}^n$, equipped with the sup-norm. Parallel to the above question of unboundedness it is natural to consider and qualify the continuity of distributions from W_p^k in dependence upon k and p. As is well-known, the counterparts of (1.1) and (1.2) yield that for $\frac{n}{p} < k < \frac{n}{p}+1$, $1 \le p < \infty$,

$$W_p^k \hookrightarrow \mathrm{Lip}^a, \quad 0 < a \le k - \frac{n}{p} < 1, \tag{1.11}$$

and,

$$W_p^{1+n/p} \hookrightarrow \mathrm{Lip}^a, \quad 0 < a < 1, \tag{1.12}$$

where Lip^a, $0 < a \le 1$, contains all $f \in C$ such that for some $c > 0$ and all $x, h \in \mathbb{R}^n$, $|f(x+h) - f(x)| \le c\,|h|^a$. Similarly to (1.2), the case $a = 1$ in (1.12) is excluded (unless $p = 1$ as some special case), i.e., there are functions from $W_p^{1+n/p}$ that are not Lipschitz-continuous. However, as some compensation, one can consider the celebrated result of Brézis and Wainger [BW80] in which it was shown that every function u in $H_p^{1+n/p}$, $1 < p < \infty$, is "almost" Lipschitz-continuous, in the sense that for all $x, y \in \mathbb{R}^n$, $0 < |x - y| < \frac{1}{2}$,

$$|u(x) - u(y)| \leq c\,|x - y|\,\left|\log|x - y|\right|^{1-\frac{1}{p}}\|u|H_p^{1+n/p}\|. \tag{1.13}$$

Here c is a constant independent of x, y and u. In [EH99] we investigated the "sharpness" of this result (concerning the exponent of the log-term), as well as possible extensions to the wider scale of F-spaces and parallel results for B-spaces. Using the classical concept of the modulus of continuity $\omega(f, t)$, (1.13) can be reformulated as

$$\sup_{0<t<1/2} \frac{\omega(f,t)}{t\,|\log t|^{1-\frac{1}{p}}} \leq c\,\|f|H_p^{1+n/p}\|, \tag{1.14}$$

which will be strengthened to

$$\left(\int_0^{\frac{1}{2}} \left[\frac{\omega(f,t)}{t\,|\log t|}\right]^p \frac{dt}{t}\right)^{1/p} \leq c\,\|f|H_p^{1+n/p}\|,$$

and an assertion similar to (1.4).

Consequently, based on observations like (1.14) we shall focus on the smoothness of functions instead of their growth; i.e., when $X \hookrightarrow C$ it makes sense to replace $f^*(t)$ by $\frac{\omega(f,t)}{t}$ in (1.5) and (1.6), and to introduce the *continuity envelope function*

$$\mathcal{E}_C^X(t) := \sup_{\|f|X\|\leq 1} \frac{\omega(f,t)}{t}, \qquad 0 < t < 1,$$

and the *continuity envelope* $\mathfrak{E}_C$ in a way completely parallel to that of $\mathcal{E}_G^X$ and $\mathfrak{E}_G$, respectively. The famous Brézis-Wainger result [BW80] then appears as

$$\mathfrak{E}_C(H_p^{1+n/p}) = \left(|\log t|^{1-\frac{1}{p}}, p\right), \qquad 1 < p < \infty,$$

whereas we obtain for Lipschitz spaces Lip^a of order $0 < a < 1$,

$$\mathfrak{E}_C(\mathrm{Lip}^a) = \left(t^{-(1-a)}, \infty\right).$$

The counterpart of (1.7) reads as

$$\mathfrak{E}_C(W_p^k) = \left(t^{-\left(\frac{n}{p}+1-k\right)}, p\right), \tag{1.15}$$

where $1 \leq p < \infty$, $k \in \mathbb{N}$, with $\frac{n}{p} < k < \frac{n}{p} + 1$, refining (1.11). Again the more elaborate results are postponed to Part II. Dealing with Besov spaces, for instance, (1.9) is complemented by

$$\mathfrak{E}_{\mathsf{C}}\left(B_{p,q}^{1+n/p}\right) = \left(|\log t|^{1-\frac{1}{q}}, q\right), \quad 1 < q \leq \infty, \quad 0 < p \leq \infty. \tag{1.16}$$

Finally, in the last two chapters we return in some sense to the beginning, now discussing different points of view, e.g., whether the envelope functions are indeed (only) *envelopes* and do not belong to the spaces themselves (posed in a suitable context); in addition, we consider assertions of a more global nature instead of local ones as before. For Sobolev spaces W_p^k, $k \in \mathbb{N}_0$, $1 \leq p < \infty$, we get

$$\mathcal{E}_{\mathsf{G}}^{W_p^k}(t) \sim t^{-\frac{1}{p}} \quad \text{for} \quad t \to \infty,$$

compared with the local assertion (1.7). Recent projects focus on the underlying domain (reaching as far as spaces defined on fractals, e.g., d-sets or h-sets in [CH06]) and inclusion of further scales of spaces. Another idea is associated with envelopes and interpolation, based on similarly constructed K-envelopes, see [Pus01]. But this is out of the scope of this book.

Some obvious applications but with surprisingly sharp results are connected with Hardy inequalities and limiting embeddings, thus coming full circle to our motivation in some sense.

In the context of compact embeddings we obtain some new and rather remarkable consequences of our preceding investigations; we combine lift arguments for envelopes, discussed before, with more abstract estimates from approximation theory and obtain results of the type

$$a_{k+1}\left(\mathrm{id}: X(U) \longrightarrow C(U)\right) \leq c\, k^{-\frac{1}{n}}\, \mathcal{E}_{\mathsf{C}}^{X}\left(k^{-\frac{1}{n}}\right), \quad k \in \mathbb{N},$$

for the approximation numbers a_k of a compact embedding $X(U) \hookrightarrow C(U)$, $U \subset \mathbb{R}^n$ being the unit ball. This describes in some cases the precise asymptotic behaviour of the approximation numbers. In combination with the above-mentioned lift arguments we obtain a counterpart in the context of growth envelopes, too.

The book is structured as follows. In Chapter 2 we recall some definitions of (classical) function spaces under consideration, and explain the situation we shall study. In Chapter 3 we introduce the growth envelope function $\mathcal{E}_{\mathsf{G}}$ and derive some of its properties. This is followed by the introduction of growth envelopes $\mathfrak{E}_{\mathsf{G}}$ in Chapter 4 also providing our main results for Lorentz-Zygmund, Sobolev and weighted (Lebesgue) spaces; Chapters 5 and 6 are parallel to Chapters 3 and 4, replacing the growth envelopes by continuity envelopes. In addition to Sobolev spaces, we explicate spaces of Lipschitz type as classical examples. It is our intention that in Part I the reader is not assumed to know more function space theory than the basics on Lebesgue (and Lorentz) spaces, Lipschitz spaces and (classical) Sobolev spaces. We describe ideas and proofs in some detail. In Part II, however, the reader should be more familiar with Fourier theory and fundamentals (like atomic decomposition techniques) of function spaces of Besov and Triebel-Lizorkin type; though we recall important definitions and characterisations as far as needed, we shall not be able to avoid some technicalities and arguments closely linked with the full generality of such scales of spaces. Those who have not yet worked in this field are recommended to special monographs for details and methods, but should be able to understand the results, which are mainly extensions of those in Part I. The last chapters contain material that might be of some interest for all again. In addition to the bibliography, there is a glossary of symbols, an index and a list of figures. The number(s) behind the symbol "▶" in the References mark the pages where the corresponding entry is quoted. Each chapter is subdivided into sections. We refer to "Chapter m, Section n" as "Section $m.n$".

At the moment it seems that those working in this field as well as the publications devoted to the subject are not yet too numerous, but permanently increasing. Though we do not pretend to give a complete survey of all that has been done so far, we shall try to report on closely linked results whenever reasonable and possible.

Chapter 2

Preliminaries, classical function spaces

Let $\mathbb{R}^n$ be Euclidean n-space and $\langle x \rangle = (2 + |x|^2)^{1/2}$, $x \in \mathbb{R}^n$. Given two (quasi-) Banach spaces X and Y, we write $X \hookrightarrow Y$ if $X \subset Y$ and the natural embedding of X in Y is continuous. All unimportant positive constants will be denoted by c, occasionally with subscripts. For some $\varkappa \in \mathbb{R}$ let

$$\varkappa_+ = \max(\varkappa, 0) \quad \text{and} \quad [\varkappa] = \max\{k \in \mathbb{Z} : k \leq \varkappa\}. \tag{2.1}$$

Moreover, for $0 < r \leq \infty$ the number r' is given by $\frac{1}{r'} := \left(1 - \frac{1}{r}\right)_+$. For non-negative functions $f, g : \mathbb{R}_+ \longrightarrow \mathbb{R}_+$, the symbol $f(t) \sim g(t)$ will mean that there are positive numbers c_1, c_2 such that for all $t > 0$,

$$c_1 \, f(t) \leq g(t) \leq c_2 \, f(t).$$

For convenience, let both dx and $|\cdot|$ stand for the (n-dimensional) Lebesgue measure ℓ_n in the sequel.

We briefly describe the concept of the non-increasing rearrangement, recall the definitions of some function spaces which will serve us as examples below. This chapter ends with a section devoted to Sobolev's famous embedding theorem.

2.1 Non-increasing rearrangements

Let $[\mathcal{R}, \mu]$ be a totally σ-finite measure space, referred to simply as *measure space* in the sequel. For a function $f : \mathcal{R} \to \mathbb{C}$, μ-measurable and finite μ-a.e., its *distribution function* $\mu_f : [0, \infty) \to [0, \infty]$ is given by

$$\mu_f(s) := \mu\left(\{x \in \mathcal{R} : |f(x)| > s\}\right), \quad s \geq 0. \tag{2.2}$$

We collect some basic properties and refer to [BS88, Prop. 2.1.3] for a proof.

Proposition 2.1 *Let $f : \mathcal{R} \to \mathbb{C}$ be a μ-measurable function, finite μ-a.e.*

(i) *The function μ_f is non-negative, decreasing, and right-continuous on $[0, \infty)$. For any $c \neq 0$,*

$$\mu_{cf}(s) = \mu_f \left(\frac{s}{|c|} \right), \quad s \geq 0.$$

(ii) *Let $g : \mathcal{R} \to \mathbb{C}$ be another μ-measurable function, finite μ-a.e. Then $|g| \leq |f| \ \mu$-a.e. implies that $\mu_g \leq \mu_f$, and*

$$\mu_{f+g}(s_1 + s_2) \leq \mu_f(s_1) + \mu_g(s_2), \quad s_1, s_2 \geq 0.$$

(iii) *Let $(f_n)_n$ be a sequence of μ-measurable functions, finite μ-a.e., such that*

$$|f| \leq \liminf_{n \to \infty} |f_n| \quad \mu\text{-a.e.,}$$

then

$$\mu_f \leq \liminf_{n \to \infty} \mu_{f_n};$$

in particular, $|f_n| \uparrow |f| \ \mu$-a.e. implies $\mu_{f_n} \uparrow \mu_f$.

We postpone an example and introduce the concept of the *non-increasing* (or *decreasing*) *rearrangement* f^* of a function f first.

Definition 2.2 *For a function $f : \mathcal{R} \to \mathbb{C}$, μ-measurable and finite μ-a.e., its* non-increasing *rearrangement* $f^* : [0, \infty) \to [0, \infty]$ *by*

$$f^*(t) = \inf \{ s \geq 0 : \mu(\{x \in \mathcal{R} : |f(x)| > s\}) \leq t \}, \quad t \geq 0. \qquad (2.3)$$

We put $\inf \emptyset = \infty$, as usual. Plainly, $f^*(t) = 0$ for $t > \mu(\mathcal{R})$.

Proposition 2.3 *Let $f : \mathcal{R} \to \mathbb{C}$ be a μ-measurable function, finite μ-a.e.*

(i) *The function f^* is non-negative, decreasing, and right-continuous on $[0, \infty)$. For any $c \neq 0$,*

$$(cf)^* (t) = |c| f^*(t), \quad t \geq 0.$$

Let $0 < r < \infty$, then

$$(|f|^r)^* (t) = (f^*(t))^r, \quad t \geq 0.$$

(ii) *Let $g : \mathcal{R} \to \mathbb{C}$ be another μ-measurable function, finite μ-a.e. Then $|g| \leq |f| \ \mu$-a.e. implies that $g^*(t) \leq f^*(t)$, $t \geq 0$, and*

$$(f + g)^* (s_1 + s_2) \leq f^*(s_1) + g^*(s_2), \quad s_1, s_2 \geq 0.$$

(iii) *Let* $(f_n)_n$ *be a sequence of μ-measurable functions, finite μ-a.e., such that*

$$|f| \leq \liminf_{n \to \infty} |f_n| \quad \mu\text{-a.e.},$$

then

$$f^*(t) \leq \liminf_{n \to \infty} f_n^*(t), \quad t \geq 0;$$

in particular, $|f_n| \uparrow |f|$ μ-a.e. implies $f_n^* \uparrow f^*$.

(iv) *When $\mu_f(s) < \infty$, then $f^*(\mu_f(s)) \leq s$; conversely, if $f^*(t) < \infty$, then $\mu_f(f^*(t)) \leq t$.*

(v) *f and f^* are equi-measurable, i.e., for $s \geq 0$,*

$$\mu_f(s) = |\{t \geq 0 : f^*(t) > s\}| = \nu_{f^*}(s), \tag{2.4}$$

where $\nu(\cdot) = |\cdot|$ stands for the usual Lebesgue measure on $\mathbb{R}_+$.

(vi) *Let $0 < p < \infty$, then*

$$\int_{\mathcal{R}} |f(x)|^p \, \mu(\mathrm{d}x) = p \int_0^\infty s^{p-1} \mu_f(s) \mathrm{d}s = \int_0^\infty f^*(t)^p \mathrm{d}t, \tag{2.5}$$

and for $p = \infty$,

$$\operatorname*{ess\,sup}_{x \in \mathcal{R}} |f(x)| = \inf\{s : \mu_f(s) = 0\} = f^*(0). \tag{2.6}$$

(vii) *Let $g : \mathcal{R} \to \mathbb{C}$ be another μ-measurable function, finite μ-a.e. Then*

$$\int_{\mathcal{R}} |f(x)| \, |g(x)| \, \mu(\mathrm{d}x) \leq \int_0^\infty f^*(t) \, g^*(t) \, \mathrm{d}t. \tag{2.7}$$

There is plenty of literature on this topic; we refer to [BS88, Ch. 2, Sect. 1], [DL93, Ch. 2, §2], and [EE04, Ch. 3], for instance. Part (vii) is well-known as the Hardy-Littlewood inequality for rearrangements.

Example 2.4 We illustrate the above concepts by a few examples. Let

$$f(x) = \sum_{j=1}^m a_j \chi_{A_j}(x),$$

where the A_j, $j = 1, \ldots, m$, are finite μ-measurable subsets of $\mathcal{R}$. Without restriction of generality we may assume that they are pairwise disjoint,

$$A_j \cap A_k = \emptyset, \quad j \neq k,$$

and that $a_1 > a_2 > \cdots > a_m > 0$.

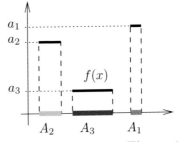

Figure 1

Clearly, $\mu_f(s) = 0$ for $s \geq a_1$, and

$$\mu_f(s) = \mu\left(\bigcup_{j=1}^k A_j\right) = \sum_{j=1}^k \mu(A_j), \quad a_{k+1} \leq s < a_k, \quad k = 1, \ldots, m,$$

where we put $a_{m+1} := 0$ for convenience. Thus we obtain

$$\mu_f(s) = \sum_{k=1}^m \left(\sum_{j=1}^k \mu(A_j)\right) \chi_{[a_{k+1}, a_k)}(s), \quad s \geq 0.$$

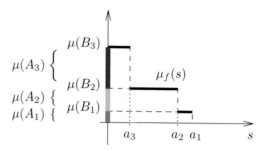

Figure 2

Using the notation $B_k := \bigcup_{j=1}^k A_j$ as in Figure 2, this can be written as

$$\mu_f(s) = \sum_{k=1}^m \mu(B_k) \chi_{[a_{k+1}, a_k)}(s), \quad s \geq 0.$$

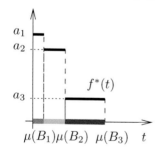

Figure 3

By (2.3) we can easily compute f^* now, since $f^*(t) = 0$, $t \geq \mu(B_m)$, and

$$f^*(t) = a_k, \quad \mu(B_{k-1}) \leq t < \mu(B_k),$$

for $k = 1, \ldots, m$, where we put $\mu(B_0) := 0$. Hence,

$$f^*(t) = \sum_{j=1}^m a_j \chi_{[\mu(B_{j-1}), \mu(B_j))}(t)$$

for $t \geq 0$.

Example 2.5 Let $[\mathcal{R}, \mu] = [\mathbb{R}_+, \nu]$, where $\nu = |\cdot|$ stands for the usual Lebesgue measure on $\mathbb{R}_+$.

We consider the function

$$g(x) = 1 - e^{-x}, \quad x > 0.$$

Then

$$\mu_g(s) = \begin{cases} \infty, & 0 \le s < 1 \\ 0, & s \ge 1 \end{cases},$$

hence

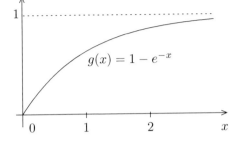

$$g^*(t) \equiv 1, \quad t \ge 0,$$

Figure 4

such that a considerable amount of information is lost.

Example 2.6 We finally consider the function ψ on $\mathbb{R}^n$,

$$\psi(x) = \begin{cases} e^{-\frac{1}{1-|x|^2}}, & |x| < 1, \\ 0, & |x| \ge 1. \end{cases} \tag{2.8}$$

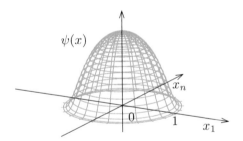

Figure 5

It is well-known that ψ is a compactly supported C^∞-function in $\mathbb{R}^n$. On the other hand one easily calculates that

$$\psi^*(t) \sim \begin{cases} e^{-\frac{1}{1-(t/|\omega_n|)^{2/n}}}, & t < |\omega_n|, \\ 0, & t \ge |\omega_n|. \end{cases} \tag{2.9}$$

Here $|\omega_n|$ denotes the (surface) measure of the unit sphere in $\mathbb{R}^n$.

Remark 2.7 For later use we recall the following lemma of Bennett and Sharpley in [BS88, Ch. 2, Lemma 2.5]: *Let* $[\mathcal{R}, \mu]$ *be a finite non-atomic measure space, and* f *a real- (or complex-) valued* μ-*measurable function* *that is finite μ-a.e. For any number s with $0 \le s \le \mu(\mathcal{R})$ there is a measurable set A_s with $\mu(A_s) = s$, such that*

$$\int_{A_s} |f| \mathrm{d}\mu \;=\; \int_0^s f^*(\tau) \mathrm{d}\tau. \tag{2.10}$$

Moreover, the sets A_s can be constructed so as to increase with s, i.e., $0 \le \sigma \le s \le \mu(\mathcal{R})$ implies $A_\sigma \subset A_s$.

2.2   Lebesgue and Lorentz spaces

Let $[\mathcal{R}, \mu]$ be a measure space again, and $0 < p \le \infty$. Then $L_p(\mathcal{R})$ are the usual (quasi-) Banach spaces consisting of all μ-a.e. finite functions f for which

$$\|f|L_p(\mathcal{R})\| = \left(\int_{\mathcal{R}} |f(x)|^p \, \mu(\mathrm{d}x) \right)^{1/p}, \tag{2.11}$$

(with the usual modification if $p = \infty$) is finite. A natural refinement of this scale of Lebesgue spaces is provided by the Lorentz (-Zygmund) spaces $L_{p,q}(\log L)_a$.

Definition 2.8 *Let $[\mathcal{R}, \mu]$ be some measure space and $0 < p, q \le \infty$.*

(i)   *The Lorentz space $L_{p,q} = L_{p,q}(\mathcal{R})$ consists of all μ-a.e. finite functions f for which the quantity*

$$\|f|L_{p,q}\| = \begin{cases} \left(\int_0^\infty \left[t^{\frac{1}{p}} f^*(t) \right]^q \frac{\mathrm{d}t}{t} \right)^{1/q} & , \quad 0 < q < \infty \\ \sup_{0<t<\infty} t^{\frac{1}{p}} f^*(t) & , \quad q = \infty \end{cases} \tag{2.12}$$

   is finite.

(ii)   *Let $a \in \mathbb{R}$. The Lorentz-Zygmund space $L_{p,q}(\log L)_a = L_{p,q}(\log L)_a(\mathcal{R})$ consists of all μ-measurable μ-a.e. finite functions f for which*

$$\|f|L_{p,q}(\log L)_a\|$$
$$= \begin{cases} \left(\int_0^\infty \left[t^{\frac{1}{p}} (1 + |\log t|)^a f^*(t) \right]^q \frac{\mathrm{d}t}{t} \right)^{1/q} & , \; q < \infty \\ \sup_{0<t<\infty} t^{\frac{1}{p}} (1 + |\log t|)^a f^*(t) & , \; q = \infty \end{cases} \tag{2.13}$$

   is finite.

Remark 2.9 These definitions are well-known and can be found, for instance, in [BS88, Ch. 4, Defs. 4.1, 6.13] and [BR80], respectively. Note that (2.12) and (2.13) do not give a norm in any case, not even for $p, q \geq 1$. However, replacing the non-increasing rearrangement f^* in (2.12) and (2.13) by its maximal function f^{**}, given by

$$f^{**}(t) = \frac{1}{t} \int_0^t f^*(s) \, ds, \quad t > 0, \tag{2.14}$$

one obtains for $1 < p < \infty$, $1 \leq q \leq \infty$, or $p = q = \infty$, a norm in that way, see [BS88, Ch. 4, Thm. 4.6]. An essential advantage of the maximal function f^{**} is that – unlike f^* – it possesses a certain sub-additivity property,

$$(f+g)^{**}(t) \leq f^{**}(t) + g^{**}(t), \quad t > 0, \tag{2.15}$$

cf. [BS88, Ch. 2, (3.10)]. Moreover, for $1 < p \leq \infty$ and $1 \leq q \leq \infty$, the corresponding expressions (2.12) with f^* and f^{**}, respectively, are equivalent; cf. [BS88, Ch. 4, Lemma 4.5].

Note that we dealt with an alternative setting of spaces $L_{p,q}(\log L)_a(\mathbb{R}^n)$ (or, more generally, when $\mu(\mathcal{R}) = \infty$) in [Har98], [Har00a]. We do not repeat this different approach based on some extrapolation procedure here in detail; the resulting spaces $L_{p,q}(\log L)_a(\mathcal{R})$ coincide with the spaces defined by (2.13) in case of a finite underlying measure space $[\mathcal{R}, \mu]$, but differ otherwise (when $\mu(\mathcal{R}) = \infty$). By our introductory remarks we had to restrict ourselves to spaces $X \subset L_1^{\mathrm{loc}}$ in general. However, we may admit in these elementary examples the full range of parameters as we stick with the case of "functions" rather than distributions at the moment.

Obviously, $L_{p,p} = L_p$, and $L_{p,q}(\log L)_0 = L_{p,q}$. Note that $L_{\infty,q}$, $0 < q < \infty$, is trivial; i.e., it contains the zero function only. The same happens for spaces of type $L_{p,q}(\log L)_a$ when $p = \infty$, $0 < q < \infty$, and $a + 1/q \geq 0$, or $p = q = \infty$, but $a > 0$. Thus when $p = \infty$ we only study spaces $L_{p,q}(\log L)_a$ in the sequel, where $a + 1/q < 0$ for $0 < q < \infty$, or $a \leq 0$ for $q = \infty$, respectively.

The spaces $L_{p,q}(\log L)_a$ are monotonically ordered in q (for fixed p and a) as well as in a (when p, q are fixed). In general, that is, when $\mu(\mathcal{R}) = \infty$, there is no monotonicity in p. But for fixed p, there is an interplay between q and a; cf. [BR80, Thms. 9.3, 9.5].

Proposition 2.10 *Let $0 < p \leq \infty$, $0 < q, r \leq \infty$, $a, b \in \mathbb{R}$, with $a + \frac{1}{q} < 0$, $b + \frac{1}{r} < 0$ if $r, q < p = \infty$, or $a \leq 0$ when $p = q = \infty$, $b \leq 0$ if $p = r = \infty$, respectively.*

(i) *Then*

$$L_{p,q}(\log L)_a \hookrightarrow L_{p,r}(\log L)_b \tag{2.16}$$

whenever

$$\text{either } q \le r, \quad a \ge b$$
$$\text{or} \quad q > r, \quad a + \frac{1}{q} > b + \frac{1}{r} \ .$$

(ii)   *Let* $0 < q \le r \le \infty$. *Then*

$$L_{\infty,q} \left(\log L\right)_a \ \hookrightarrow \ L_{\infty,r} \left(\log L\right)_b \tag{2.17}$$

if

$$a + \frac{1}{q} \ = \ b + \frac{1}{r} \ .$$

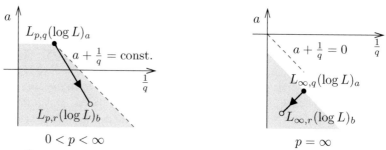

Figure 6

These conditions can, in general, not be relaxed, see [BR80, Rem. 9.4]. According to (ii), spaces $L_{\infty,q} \left(\log L\right)_a$ are ordered along the "diagonals" $a + 1/q = $ const., see also Figure 6 where we indicated in the shaded areas admitted parameters $(\frac{1}{r}, b)$ for target spaces $L_{p,r} \left(\log L\right)_b$ such that for a fixed source space we have $L_{p,q}(\log L)_a \hookrightarrow L_{p,r} \left(\log L\right)_b$.

For finite measure spaces $[\mathcal{R}, \mu]$, say, $\mu(\mathcal{R}) = 1$, and for $p = q = \infty$ and $a \ge 0$, one has $L_{\infty,\infty} \left(\log L\right)_{-a} = L_{\exp,a}$, where the latter are the *Zygmund spaces* consisting of all μ-measurable functions f on $\mathcal{R}$ for which there is a constant $\lambda = \lambda(f) > 0$ such that

$$\int_{\mathcal{R}} \exp\left(\lambda |f(x)|\right)^{1/a} d\mu \ < \ \infty, \tag{2.18}$$

(if $a = 0$, this is interpreted as f is bounded, i.e., $L_{\exp,0} = L_\infty$). In particular, for $a \ge 0$,

$$\|f | L_{\exp,a}\| \ \sim \ \sup_{0<t<1} \left(1 - \log t\right)^{-a} f^*(t), \tag{2.19}$$

cf. [BS88, Ch. 4, Def. 6.11, Lemma 6.12].

The spaces L_p, $1 \leq p \leq \infty$, belong to the category of *Banach function spaces* (or *lattices*); we briefly recall this notion and follow [BS88, Ch. 1, Sect. 1] closely. Let $[\mathcal{R}, \mu]$ be a measure space and $\mathcal{M}^+$ the cone of μ-measurable functions on $\mathcal{R}$ with non-negative values. The characteristic function of a μ-measurable set A is denoted by χ_A.

Definition 2.11 *A mapping* $\varrho : \mathcal{M}^+ \to [0, \infty]$ *is called a* Banach function quasi-norm *if, for all* g, f, f_n, $n \in \mathbb{N}$, *in* $\mathcal{M}^+$, *for all* $\gamma \geq 0$, *and for all* μ-measurable subsets A of $\mathcal{R}$, *the following properties hold:*

(i) $\varrho(f) = 0$ *if, and only if,* $f = 0$ μ-a.e.

(ii) $\varrho(\gamma f) = \gamma\,\varrho(f)$

(iii) $\varrho(f + g) \leq c_\varrho\Big(\varrho(f) + \varrho(g) \Big)$ *for some* $c_\varrho \geq 1$

(iv) $0 \leq g \leq f$ μ-a.e. implies $\varrho(g) \leq \varrho(f)$

(v) $0 \leq f_n \uparrow f$ μ-a.e. implies $\varrho(f_n) \uparrow \varrho(f)$

(vi) $\mu(A) < \infty$ *implies* $\varrho(\chi_A) < \infty$

(vii) $\mu(A) < \infty$ *implies* $\displaystyle\int_A f\mathrm{d}\mu \leq c_{A,\varrho}\,\varrho(f)$ *for some constant* $c_{A,\varrho}$
          depending on A *and* ϱ, *but independent of* f

Note that a mapping $\varrho : \mathcal{M}^+ \to [0, \infty]$ is called a *Banach function norm* (or simply *function norm*) if it satisfies conditions (i)-(vii) with $c_\varrho = 1$ in (iii).

Definition 2.12 *Let* ϱ *be a function quasi-norm over* $[\mathcal{R}, \mu]$. *The collection* $X = X(\varrho)$ *of all real or complex valued* μ-measurable functions on $\mathcal{R}$, *for which* $\varrho(|f|) < \infty$, *is called a* quasi-Banach function space *over* $[\mathcal{R}, \mu]$. *The quasi-norm of a function* $f \in X$ *is given by* $\|f|X\| = \varrho(|f|)$.

If ϱ is a Banach function norm, then $\|\cdot|X\|$ is a norm, X is called a *Banach function space*, and X endowed with the norm $\|\cdot|X\|$ is a Banach space. Returning to our above-explained notation $X \hookrightarrow Y$ let us remark that for Banach function spaces X and Y over the same measure space $[\mathcal{R}, \mu]$ the condition $X \subset Y$ already implies $X \hookrightarrow Y$; cf. [BS88, Ch. 1, Thms. 1.6, 1.8].

2.3   Spaces of continuous functions

Let $C(\mathbb{R}^n)$ be the space of all complex-valued bounded uniformly continuous functions on $\mathbb{R}^n$, equipped with the sup-norm as usual. If $m \in \mathbb{N}$, we define $C^m(\mathbb{R}^n) = \{f : D^\alpha f \in C(\mathbb{R}^n) \text{ for all } |\alpha| \le m\}$. Here $\alpha = (\alpha_1, \ldots, \alpha_n) \in \mathbb{N}_0^n$ stands for some multi-index,

$$|\alpha| = \alpha_1 + \cdots + \alpha_n, \quad \alpha \in \mathbb{N}_0^n,$$

D^α are classical derivatives,

$$D^\alpha = \frac{\partial^{|\alpha|}}{\partial x_1^{\alpha_1} \ldots \partial x_n^{\alpha_n}}, \quad \alpha \in \mathbb{N}_0^n,$$

and $C^m(\mathbb{R}^n)$ is endowed with the norm

$$\|f|C^m(\mathbb{R}^n)\| = \sum_{|\alpha| \le m} \|D^\alpha f|C(\mathbb{R}^n)\|,$$

where $\| \cdot |C(\mathbb{R}^n)\|$ can obviously be replaced by $\| \cdot |L_\infty(\mathbb{R}^n)\|$ in this case. The set of all compactly supported, infinitely often differentiable functions is denoted by $C_0^\infty(\mathbb{R}^n)$, as usual.

Recall the concept of the difference operator Δ_h^m, $m \in \mathbb{N}_0$, $h \in \mathbb{R}^n$: Let f be an arbitrary function on $\mathbb{R}^n$; then

$$(\Delta_h^1 f)(x) = f(x + h) - f(x), \quad (\Delta_h^{m+1} f)(x) = \Delta_h^1(\Delta_h^m f)(x), \quad (2.20)$$

where $x, h \in \mathbb{R}^n$. For convenience we may write Δ_h instead of Δ_h^1. Furthermore, the *r-th modulus of smoothness* of a function $f \in L_p(\mathbb{R}^n)$, $1 \le p \le \infty$, $r \in \mathbb{N}$, is defined by

$$\omega_r(f, t)_p = \sup_{|h| \le t} \|\Delta_h^r f|L_p(\mathbb{R}^n)\|, \quad t > 0, \quad (2.21)$$

see [BS88, Ch. 5, Def. 4.2] or [DL93, Ch. 2, §7]. Note that each modulus $\omega_r(f, t)_p$, $1 \le p \le \infty$, $r \in \mathbb{N}$, is a non-negative, continuous, increasing function of $t > 0$. Obviously,

$$\omega_r(f, t)_p \le 2^r \|f|L_p(\mathbb{R}^n)\|, \quad r \in \mathbb{N}, \quad 1 \le p \le \infty, \quad t > 0.$$

Moreover, $\omega_r(f, t)_p \searrow \omega_r(f, 0)_p = 0$ for $t \downarrow 0$. We shall write $\omega(f, t)_p$ instead of $\omega_1(f, t)_p$ and omit the index $p = \infty$ if there is no danger of confusion, that is, $\omega(g, t)$ instead of $\omega(g, t)_\infty$. We refer to the literature mentioned above for further details.

Marchaud's inequality states the following: let $f \in L_p(\mathbb{R}^n)$, $1 \leq p \leq \infty$, $t > 0$, and $k \in \mathbb{N}$; then

$$\omega_k(f,t)_p \leq \frac{k}{\log 2} t^k \int_t^\infty \frac{\omega_{k+1}(f,u)_p}{u^k} \frac{du}{u}, \qquad (2.22)$$

see [BS88, Ch. 5, (4.11)] or [DL93, Ch. 2, Thm. 8.1] (for the one-dimensional case).

When $s \in \mathbb{R}_+$ we consider Hölder-Zygmund spaces $\mathcal{C}^s$. For a positive number $a \in \mathbb{R}_+$, let

$$a = \lfloor a \rfloor + \{a\}, \quad \lfloor a \rfloor := \max\{k \in \mathbb{Z} : k < a\}, \quad 0 < \{a\} \leq 1. \qquad (2.23)$$

Definition 2.13 *Let $s > 0$. The Hölder-Zygmund space $\mathcal{C}^s(\mathbb{R}^n)$ consists of all $f \in C^{\lfloor s \rfloor}(\mathbb{R}^n)$, such that*

$$\|f|\mathcal{C}^s(\mathbb{R}^n)\| = \left\|f|C^{\lfloor s \rfloor}(\mathbb{R}^n)\right\| + \sum_{|\alpha|=\lfloor s \rfloor} \sup_{h \neq 0} \frac{\|\Delta_h^2 D^\alpha f|C(\mathbb{R}^n)\|}{|h|^{\{s\}}} \qquad (2.24)$$

is finite.

It is well-known that $C^m(\mathbb{R}^n) \neq \mathcal{C}^m(\mathbb{R}^n)$, $m \in \mathbb{N}$. However, these Hölder-Zygmund spaces $\mathcal{C}^s$ fit precisely in the scale of Besov spaces that we study in Part II.

Finally, as for smoothness assertions of order 1, we shall be concerned with Lipschitz spaces, too.

Definition 2.14 *Let $0 < a \leq 1$. The Lipschitz space $\mathrm{Lip}^a(\mathbb{R}^n)$ is defined as the set of all $f \in C(\mathbb{R}^n)$ such that*

$$\|f|\mathrm{Lip}^a(\mathbb{R}^n)\| := \|f|C(\mathbb{R}^n)\| + \sup_{0<t<1} \frac{\omega(f,t)}{t^a} \qquad (2.25)$$

is finite.

Remark 2.15 Note that the restriction $0 < a \leq 1$ is quite natural, as otherwise the spaces contain only constants; when $a = 1$ one recovers the classical Lipschitz space $\mathrm{Lip}^1(\mathbb{R}^n)$, normed by

$$\left\|f|\mathrm{Lip}^1(\mathbb{R}^n)\right\| = \|f|C(\mathbb{R}^n)\| + \sup_{0<t<1} \frac{\omega(f,t)}{t}. \qquad (2.26)$$

If $0 < a < 1$, then $\mathrm{Lip}^a(\mathbb{R}^n) = \mathcal{C}^a(\mathbb{R}^n)$.

For later use we shall introduce some refinement of the scale of Lipschitz spaces. This essentially follows the ideas presented in [EH99], [EH00], [Har00b].

Definition 2.16 *Let* $1 \le p \le \infty$, $0 < q \le \infty$, $\alpha > \frac{1}{q}$ *(with* $\alpha \ge 0$ *if* $q = \infty$). *Then* $\text{Lip}_{p,q}^{(1,-\alpha)}(\mathbb{R}^n)$ *is defined as the set of all* $f \in L_p(\mathbb{R}^n)$ *such that*

$$\left\| f | \text{Lip}_{p,q}^{(1,-\alpha)}(\mathbb{R}^n) \right\| := \| f | L_p(\mathbb{R}^n) \| + \left(\int_0^{\frac{1}{2}} \left[\frac{\omega(f,t)_p}{t \, | \log t|^\alpha} \right]^q \frac{dt}{t} \right)^{1/q} \qquad (2.27)$$

(with the usual modification if $q = \infty$) *is finite.*

Note that Definition 2.16 coincides with [EH00, Def. 4.1] when $q = \infty$, and in case of $p = q = \infty$, $\alpha \ge 0$, we recover the logarithmic Lipschitz spaces, $\text{Lip}^{(1,-\alpha)} = \text{Lip}_{\infty,\infty}^{(1,-\alpha)}$ introduced in [EH99, Def. 1.1]. In particular, for $\alpha = 0$, $p = q = \infty$, these are nothing else than the classical Lipschitz spaces $\text{Lip}^1(\mathbb{R}^n)$.

Convention.   As long as there is no danger of confusion we shall omit the "∞"-indices in the classical setting, i.e., we write $\text{Lip}^{(1,-\alpha)}$ instead of $\text{Lip}_{\infty,\infty}^{(1,-\alpha)}$.

The restriction $\alpha > \frac{1}{q}$ is quite natural as otherwise we have $\text{Lip}_{p,q}^{(1,-\alpha)} = \{0\}$ only, see [Har00b, Rem. 18]. However, when $q = \infty$ we may also admit $\alpha = 0$, whereas $\text{Lip}^{(1,-\alpha)}$ would consist only of constants were α allowed to be negative. The somehow unusual notation using $-\alpha$ (instead of α) is simply due to the fact that we want to emphasise that the additional smoothness parameter α acts in such a way that the usual spaces $\text{Lip}^1(\mathbb{R}^n)$ are extended: $\text{Lip}^1(\mathbb{R}^n) \hookrightarrow \text{Lip}^{(1,-\alpha)}(\mathbb{R}^n)$ for all $\alpha \ge 0$, i.e., the spaces become larger when less smoothness is assumed – as it should be in some reasonable notation.

Remark 2.17 The spaces $\text{Lip}^{(1,-\alpha)}(\mathbb{R}^n)$, $\alpha \ge 0$, can also be obtained as a special case of the more general spaces $C^{0,\sigma(t)}(\overline{\Omega})$, $\Omega \subseteq \mathbb{R}^n$, which were introduced by Kufner, John and Fučík; see [KJF77, Def. 7.2.12]. Moreover, spaces of type $\text{Lip}_{p,\infty}^{(1,0)} = \text{Lip}(1,L_p)$ are considered by DeVore and Lorentz in [DL93, Ch. 2, §9], where $\mathbb{R}^n$ is being replaced by some interval $[a,b] \subset \mathbb{R}$ and $0 < p \le \infty$. Similarly, spaces $\text{Lip}(\alpha,p)$ were studied by Kolyada in [Kol89].

We introduce the *Zygmund spaces* $C^{(1,-\alpha)}(\mathbb{R}^n)$, $\alpha \ge 0$, as refinements of $C^1(\mathbb{R}^n)$, given by Definition 2.13, and counterparts of the spaces $\text{Lip}^{(1,-\alpha)}$, see also [EH99].

Definition 2.18 *Let* $\alpha \geq 0$. *Then the space* $\mathcal{C}^{(1,-\alpha)}(\mathbb{R}^n)$ *is defined as the set of all* $f \in C(\mathbb{R}^n)$ *such that*

$$\|f|\mathcal{C}^{(1,-\alpha)}(\mathbb{R}^n)\| = \|f|L_\infty(\mathbb{R}^n)\| + \sup_{\substack{x,h \in \mathbb{R}^n \\ 0 < |h| < 1/2}} \frac{|(\Delta_h^2 f)(x)|}{|h| \, |\log|h||^\alpha} < \infty. \qquad (2.28)$$

Remark 2.19 Though it might not be obvious at first glance there is an essential difference between spaces of type, say, $\mathrm{Lip}^{(1,-\alpha)}$ and $\mathcal{C}^{(1,-\alpha)}$, $\alpha \geq 0$ – concerning their compatibility with spaces of Besov type as will be clear in Part II, especially in Section 7.2.

Remark 2.20 In view of applications, suitably adapted Hölder inequalities are often needed; we give an example for spaces $\mathrm{Lip}_{p,\infty}^{(1,-\alpha)}$, that can be found in [EH00, Prop. 4.3, Rem. 4.4]. Let $1 \leq p,q \leq \infty$ such that $0 \leq \frac{1}{r} = \frac{1}{p} + \frac{1}{q} \leq 1$. Let $\alpha, \beta \geq 0$. Then

$$\mathrm{Lip}_{p,\infty}^{(1,-\alpha)} \cdot \mathrm{Lip}_{q,\infty}^{(1,-\beta)} \hookrightarrow \mathrm{Lip}_{r,\infty}^{(1,-\max(\alpha,\beta))} \hookrightarrow \mathrm{Lip}_{r,\infty}^{(1,-(\alpha+\beta))}. \qquad (2.29)$$

The following extrapolation type result for spaces $\mathrm{Lip}_{p,q}^{(1,-\alpha)}$ was obtained in [EH00, Prop. 4.2(i)], [Har00b, Prop. 7]; for details about extrapolation techniques we refer to [Mil94].

Proposition 2.21 *Let* $1 \leq p \leq \infty$.

(i) *Let* $q = \infty$, $\alpha > 0$. *Then* $f \in \mathrm{Lip}_{p,\infty}^{(1,-\alpha)}$ *if, and only if,* f *belongs to* L_p *and there is some* $c > 0$ *such that for all* λ, $0 < \lambda < 1$,

$$\sup_{0 < t < 1/2} \frac{\omega(f,t)_p}{t^{1-\lambda}} \leq c\,\lambda^{-\alpha}.$$

Moreover, we obtain as an equivalent norm in $\mathrm{Lip}_{p,\infty}^{(1,-\alpha)}$,

$$\left\|f|\mathrm{Lip}_{p,\infty}^{(1,-\alpha)}\right\| \sim \|f|L_p\| + \sup_{0 < \lambda < 1} \lambda^\alpha \sup_{0 < t < 1/2} \frac{\omega(f,t)_p}{t^{1-\lambda}}. \qquad (2.30)$$

(ii) *Let* $0 < q < \infty$, $\alpha > \frac{1}{q}$. *Then* $f \in \mathrm{Lip}_{p,q}^{(1,-\alpha)}$ *if, and only if,* f *belongs to* L_p *and there is some* $c > 0$ *such that*

$$\int_0^1 \lambda^{\alpha q} \int_0^{\frac{1}{2}} \left[\frac{\omega(f,t)_p}{t^{1-\lambda}}\right]^q \frac{\mathrm{d}t}{t} \frac{\mathrm{d}\lambda}{\lambda} \leq c.$$

Moreover,

$$\left\| f | \mathrm{Lip}_{p,\,q}^{(1,-\alpha)} \right\| \sim \| f | L_p \| + \left(\int_0^1 \lambda^{\alpha q} \int_0^{\frac{1}{2}} \left[\frac{\omega(f,t)_p}{t^{1-\lambda}} \right]^q \frac{dt}{t} \frac{d\lambda}{\lambda} \right)^{1/q}. \quad (2.31)$$

Remark 2.22 When $p = \infty$ Proposition 2.21(i) coincides with the result of Krbec and Schmeisser in [KS01, Prop. 2.5] which was also our motivation for the above extension; part (i) was already presented in [EH00, Prop. 4.2(i)].

We give a (sharp) embedding result for logarithmic Lipschitz spaces $\mathrm{Lip}^{(1,-\alpha)}$ and Zygmund spaces $\mathcal{C}^{(1,-\alpha)}$ in the spirit of Proposition 2.10.

Proposition 2.23 *Let* $1 \le p \le \infty$, $0 < q, r \le \infty$, $\alpha > \frac{1}{q}$, $\beta > \frac{1}{r}$. *Then*

$$\mathrm{Lip}_{p,\,q}^{(1,-\alpha)} \hookrightarrow \mathrm{Lip}_{p,\,r}^{(1,-\beta)} \quad (2.32)$$

if, and only if,

$$\text{either} \quad r \ge q, \quad \beta - \frac{1}{r} \ge \alpha - \frac{1}{q},$$

$$\text{or} \quad r < q, \quad \beta - \frac{1}{r} > \alpha - \frac{1}{q}.$$

Remark 2.24 The result was proved in [Har00b, Prop. 16]. The similarity to Proposition 2.10 is obvious. Let us again point out the somehow astonishing result that concerning the embedding $\mathrm{Lip}_{p,\,q}^{(1,-\alpha)}$ into $\mathrm{Lip}_{p,\,r}^{(1,-\beta)}$ one can "compensate" some gain of logarithmic smoothness $-\beta > -\alpha$ by "paying" with the additional index q, that is, as long as $(-\beta) - (-\alpha) \le \frac{1}{q} - \frac{1}{r}$, $r \ge q$.

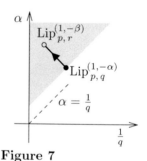

Figure 7

Dealing with Zygmund spaces $\mathcal{C}^{(1,-\alpha)}$, we restrict ourselves to $p = q = \infty$ at the moment (but will return to this setting in a more general approach in Section 7.2. Recall our convention $\mathrm{Lip}^{(1,-\alpha)} = \mathrm{Lip}_{\infty,\infty}^{(1,-\alpha)}$. We obtained in [EH00, Prop. 2.7] the following embedding result.

Proposition 2.25 *Let* α, β, γ *be non-negative real numbers. Then*

$$\mathrm{Lip}^{(1,-\alpha)} \hookrightarrow \mathcal{C}^{(1,-\beta)} \hookrightarrow \mathrm{Lip}^{(1,-\gamma)} \quad (2.33)$$

if, and only if,

$$\beta \geq \alpha, \quad and \quad \gamma \geq \beta + 1.$$

Finally we consider spaces $\mathrm{Lip}_{\infty, q}^{(a, -\alpha)}$, $0 < a \leq 1$, $0 < q \leq \infty$, $\alpha \in \mathbb{R}$, which are "close" to Lip^a. For convenience, we deal with $p = \infty$ exclusively.

Definition 2.26 *Let* $0 < a < 1$, $0 < q \leq \infty$, *and*

$$\begin{cases} \alpha \in \mathbb{R} & if \quad 0 < a < 1, 0 < q \leq \infty \\ \alpha > \frac{1}{q} & if \quad a = 1, 0 < q < \infty \\ \alpha \geq 0 & if \quad a = 1, \quad q = \infty \end{cases}. \tag{2.34}$$

The space $\mathrm{Lip}_{\infty, q}^{(a, -\alpha)}(\mathbb{R}^n)$ *is defined as the set of all* $f \in C(\mathbb{R}^n)$ *such that*

$$\left\| f | \mathrm{Lip}_{\infty, q}^{(a, -\alpha)}(\mathbb{R}^n) \right\| := \| f | C(\mathbb{R}^n) \| + \left(\int_0^{\frac{1}{2}} \left[\frac{\omega(f, t)}{t^a |\log t|^\alpha} \right]^q \frac{dt}{t} \right)^{1/q} \tag{2.35}$$

(with the usual modification if $q = \infty$) is finite.

Remark 2.27 The above spaces first appeared (in this notation) in [Har00b] in connection with limiting embeddings, extending the case $a = 1$ studied in [EH99] and [EH00]. We obtained different characterisations of spaces $\mathrm{Lip}_{\infty, q}^{(a, -\alpha)}$ in the sense of Proposition 2.21: Let $0 < a \leq 1$, $0 < q \leq \infty$, and $\alpha \in \mathbb{R}$ satisfy (2.34). Then $f \in \mathrm{Lip}_{\infty, q}^{(a, -\alpha)}$ if, and only if, f belongs to C and there is some $c > 0$ such that

$$\int_0^a \lambda^{\alpha q} \int_0^{\frac{1}{2}} \left[\frac{\omega(f, t)}{t^{a-\lambda}} \right]^q \frac{dt}{t} \frac{d\lambda}{\lambda} \leq c.$$

Moreover,

$$\left\| f | \mathrm{Lip}_{\infty, q}^{(a, -\alpha)} \right\| \sim \| f | C \| + \left(\int_0^a \lambda^{\alpha q} \int_0^{\frac{1}{2}} \left[\frac{\omega(f, t)}{t^{a-\lambda}} \right]^q \frac{dt}{t} \frac{d\lambda}{\lambda} \right)^{1/q}. \tag{2.36}$$

Let us finally mention that there are interesting applications connected with Lipschitz spaces $\mathrm{Lip}_{\infty, q}^{(1, -\alpha)}$, see the book [Lio98] by Lions and the paper [Vis98] by Vishik.

2.4   Sobolev spaces, Sobolev's embedding theorem

Let $W_p^k(\mathbb{R}^n)$ be the space of those functions f on $\mathbb{R}^n$ (or locally regular distributions $f \in L_1^{\text{loc}}(\mathbb{R}^n)$) for which all weak derivatives of order at most $k \in \mathbb{N}$, $D^\alpha f$, $|\alpha| \leq k$, belong to $L_p(\mathbb{R}^n)$, $1 \leq p < \infty$. It is well-known that the space $W_p^k(\mathbb{R}^n)$ is a Banach space equipped with the norm

$$\|f|W_p^k(\mathbb{R}^n)\| = \Big(\sum_{|\alpha|\leq k} \|D^\alpha f|L_p(\mathbb{R}^n)\|^p \Big)^{1/p}. \tag{2.37}$$

Moreover, the compactly supported smooth functions, $C_0^\infty(\mathbb{R}^n)$, are dense in $W_p^k(\mathbb{R}^n)$. There is a great deal of literature on Sobolev spaces, see [AF03], [Maz85], [EE87], [Zie89]. Dealing with Sobolev spaces on domains, $W_p^k(\Omega)$, questions of extendability, definition by restriction procedures or – alternatively – by intrinsic characterisations, connected with the geometry of the underlying domain, too, are highly non-trivial and well-studied. In the present book we are, however, essentially interested in spaces on $\mathbb{R}^n$ and will thus postpone a more refined discussion of spaces on domains to elsewhere.

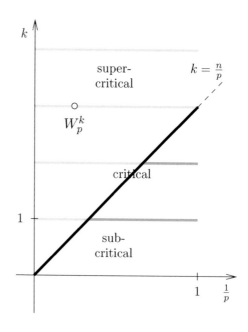

Figure 8

We recall the famous Sobolev embedding theorem in a version used in the sequel. Note, in particular, that we consider here the case $\Omega = \mathbb{R}^n$ only, though there are many extensions for (bounded) domains $\Omega \subset \mathbb{R}^n$ that satisfy certain conditions (cone condition, segment condition, local Lipschitz condition, etc.). We rely here on the formulation given in [AF03, Thm. 4.12] and [EE87, Ch. V] with $\Omega = \mathbb{R}^n$. In the above $(\frac{1}{p}, s)$-diagram we indicate the different cases that will be considered later on in detail. Each space W_p^k is marked here by its smoothness parameter $s = k \in \mathbb{N}_0$ and the integrability $p \in [1, \infty)$. This diagram will be enriched and filled when we deal with spaces of type $B_{p,q}^s$ and $F_{p,q}^s$ in Part II. As we deal with spaces on $\mathbb{R}^n$ only, we shall omit the "$\mathbb{R}^n$" from their notation as long as there is no danger of confusion.

Theorem 2.28 *Let* $k \in \mathbb{N}$ *and* $1 \le p < \infty$.

(i)   super-critical case

   Assume that $k > \dfrac{n}{p}$ *or* $k = n$ *and* $p = 1$; *then*

$$W_p^k \hookrightarrow C, \qquad (2.38)$$

and

$$W_p^{k+1} \hookrightarrow \mathrm{Lip}^1. \qquad (2.39)$$

Moreover, for $p \le r \le \infty$,

$$W_p^k \hookrightarrow L_r. \qquad (2.40)$$

If, in addition, $k < \dfrac{n}{p} + 1$, *then*

$$W_p^k \hookrightarrow \mathrm{Lip}^a, \quad 0 < a \le k - \frac{n}{p} < 1, \qquad (2.41)$$

and, for $k = \dfrac{n}{p} + 1$,

$$W_p^{1+n/p} \hookrightarrow \mathrm{Lip}^a, \quad 0 < a < 1, \qquad (2.42)$$

with

$$W_1^{1+n} \hookrightarrow \mathrm{Lip}^a, \quad 0 < a \le 1. \qquad (2.43)$$

(ii)   critical case

   Assume that $k = \dfrac{n}{p}$, *then*

$$W_p^k \hookrightarrow L_r \qquad (2.44)$$

for $p \le r < \infty$.

(iii) <u>sub-critical case</u>

  Assume that $k < \dfrac{n}{p}$, *then*

$$W_p^k \hookrightarrow L_r \tag{2.45}$$

  for $p \le r \le p^* = \dfrac{np}{n - kp}$.

P r o o f: As mentioned above, there are many proofs in the literature. For convenience, we give a proof here that essentially follows [AF03, Thm. 4.12] and [EE87, Ch. V] and deal with $\Omega = \mathbb{R}^n$ only.

  <u>Step 1.</u>   We start with the sub-critical case (iii) and $k = 1$; thus assume $1 \le p < n$, and let p^* be given by $p^* = \frac{np}{n-p}$. First we show that there exists a constant $c > 0$ such that for all $u \in W_p^1$,

$$\|u|L_{p^*}\| \le c \, \|u|W_p^1\| . \tag{2.46}$$

By the density of C_0^∞ in W_p^k we may assume by standard arguments that $u \in C_0^1 \cap W_p^1$. Let $\alpha = (\alpha_1, \ldots, \alpha_n) \in \mathbb{N}_0^n$, $|\alpha| = 1$, then $\alpha = \alpha^{(i)} = (0, \ldots, 1, \ldots, 0)$ with $\alpha_k = \delta_{ik}$, $i, k = 1, \ldots, n$. We write in obvious notation D_i for $D^{\alpha^{(i)}}$, $i = 1, \ldots, n$. Let $x = (x_1, \ldots, x_n) \in \mathbb{R}^n$, then

$$u(x) = \int\limits_{-\infty}^{x_1} D_1 u(\xi_1, x_2, \ldots, x_n) d\xi_1 = - \int\limits_{x_1}^{\infty} D_1 u(\xi_1, x_2, \ldots, x_n) d\xi_1,$$

hence

$$|u(x)| \le \frac{1}{2} \int\limits_{-\infty}^{\infty} |D_1 u(\xi_1, x_2, \ldots, x_n)| \, d\xi_1,$$

and, dealing with the other coordinates in a similar way,

$$|2u(x)|^n \le \prod_{j=1}^{n} \left(\int\limits_{-\infty}^{\infty} |D_j u(x_1, \ldots, \xi_j, \ldots, x_n)| \, d\xi_j \right) . \tag{2.47}$$

We raise (2.47) to the power $\frac{1}{n-1}$, integrate it with respect to x_1 and apply an iterated version of Hölder's inequality in the form

$$\int\limits_{-\infty}^{\infty} \left(\prod_{k=1}^{n-1} g_k(t) \right)^{\frac{1}{n-1}} dt \le \prod_{k=1}^{n-1} \left(\int\limits_{-\infty}^{\infty} g_k(t) dt \right)^{\frac{1}{n-1}} ;$$

this gives

$$\int_{-\infty}^{\infty} |2u(x_1,\ldots,x_n)|^{\frac{n}{n-1}}\,dx_1$$

$$\leq \left(\int_{-\infty}^{\infty} |D_1 u(x)|\,dx_1\right)^{\frac{1}{n-1}} \prod_{k=2}^{n}\left(\int_{-\infty}^{\infty}\int_{-\infty}^{\infty} |D_k u(x)|\,dx_k dx_1\right)^{\frac{1}{n-1}}.$$

We proceed with the remaining coordinates in the same way, where every time precisely $n-1$ factors are specifically involved in the integration; thus

$$\int_{\mathbb{R}^n} |2u(x)|^{\frac{n}{n-1}}\,dx \leq \prod_{k=1}^{n}\left(\int_{\mathbb{R}^n} |D_k u(x)|\,dx\right)^{\frac{1}{n-1}},$$

which can be reformulated as

$$\left\|u|L_{\frac{n}{n-1}}\right\| \leq \frac{1}{2}\prod_{k=1}^{n}\|D_k u|L_1\|^{\frac{1}{n}}. \tag{2.48}$$

The arithmetic-geometric inequality implies

$$\left\|u|L_{\frac{n}{n-1}}\right\| \leq \frac{1}{2n}\sum_{k=1}^{n}\|D_k u|L_1\| \leq \frac{1}{2n}\left\|u|W_1^1\right\|, \tag{2.49}$$

which is the desired result for $p=1$. When $p>1$, then $\frac{(n-1)p}{n-p}>1$, and we apply (2.49) to $v=|u|^{\frac{(n-1)p}{n-p}}$ which satisfies the required differentiability conditions. Consequently,

$$\|u|L_{p^*}\| = \left\|v|L_{\frac{n}{n-1}}\right\|^{\frac{n-p}{(n-1)p}}$$

$$\leq c_1 \left(\sum_{k=1}^{n}\|D_k v|L_1\|\right)^{\frac{n-p}{(n-1)p}}$$

$$\leq c_2 \left(\sum_{k=1}^{n}\int_{\mathbb{R}^n} |u(x)|^{\frac{n(p-1)}{n-p}} |D_k u(x)|\,dx\right)^{\frac{n-p}{(n-1)p}}$$

$$\leq c_3 \|u|L_{p^*}\|^{\frac{n(p-1)}{p(n-1)}}\left(\sum_{k=1}^{n}\|D_k u|L_p\|\right)^{\frac{n-p}{(n-1)p}}$$

$$\leq c_4 \|u|L_{p^*}\|^{\frac{n(p-1)}{p(n-1)}}\left\|u|W_p^1\right\|^{\frac{n-p}{(n-1)p}},$$

where we used Hölder's inequality again. This leads to (2.46).

Step 2.   We complete the proof of (iii). Assume $k \in \mathbb{N}$, $k < \frac{n}{p}$, and $u \in W_p^k$. Consider p_j given by $p_{j+1} = \frac{np_j}{n-p_j}$, $j = 0, \ldots, k-1$, with $p_0 = p$; thus $p_k = p^*$. We apply (2.46) and obtain consecutively

$$\|u|L_{p^*}\| \le c_1 \sum_{|\alpha|=1} \|D^\alpha u|L_{p_{k-1}}\| \le c_2 \sum_{|\beta|=2} \|D^\beta u|L_{p_{k-2}}\|$$

$$\vdots$$

$$\le c_k \sum_{|\gamma|=k} \|D^\gamma u|L_{p_0}\| \le c \|u|W_p^k\|. \qquad (2.50)$$

Thus (2.45) is verified for $r = p^*$ and, simply by definition, for $r = p$. Let now r be such that $p < r < p^*$, and $u \in W_p^k$; then another application of Hölder's inequality together with (2.50) leads to

$$\|u|L_r\| = \left(\int_{\mathbb{R}^n} |u(x)|^{\frac{p(p^*-r)}{p^*-p}} |u(x)|^{\frac{p^*(r-p)}{p^*-p}} \, dx \right)^{\frac{1}{r}}$$

$$\le \|u|L_p\|^{\frac{p(p^*-r)}{r(p^*-p)}} \|u|L_{p^*}\|^{\frac{p^*(r-p)}{r(p^*-p)}}$$

$$\le c \|u|W_p^k\|^{\frac{p(p^*-r)}{r(p^*-p)}} \|u|W_p^k\|^{\frac{p^*(r-p)}{r(p^*-p)}} = c \|u|W_p^k\|.$$

Step 3.   We study the limiting case $k = \frac{n}{p}$. We may assume $p > 1$, the rest is postponed to Step 4. First let $u \in W_p^{n/p}$ with supp $u \subset \Omega$, $|\Omega| < \infty$, and $r \ge p' = \frac{p}{p-1}$. Then we have for $s := \frac{pr}{p+r}$ that $1 \le s < p$ and $r = \frac{ns}{n-ks} = s^*$. By Hölder's inequality we obtain that

$$\|u|W_s^k\| \le c|\Omega|^{\frac{1}{s}-\frac{1}{p}} \|u|W_p^k\|.$$

Moreover, we may apply (iii) to W_s^k and L_r since $k = \frac{n}{p} \frac{pr}{s(p+r)} = \frac{n}{s} \frac{r}{p+r} < \frac{n}{s}$, $r = s^*$, leading to

$$\|u|L_r\| \le c \|u|W_s^{n/p}\| \le c' |\Omega|^{\frac{1}{r}} \|u|W_p^{n/p}\|. \qquad (2.51)$$

When $p < r < p'$, hence $1 < p < 2$, then similar to Step 2 we may interpolate (2.51) and $W_p^k \hookrightarrow L_p$; let $\varkappa := \frac{p'-p}{p'-r} > 1$, then by Hölder's inequality again together with (2.51) for $r = p'$,

$$\|u|L_r\|^r = \int\limits_{\mathbb{R}^n} |u(x)|^{\frac{p}{r}} \, |u(x)|^{\frac{p'}{r'}} \, dx$$

$$\leq \|u|L_p\|^{\frac{p}{r}} \, \|u|L_{p'}\|^{\frac{p'}{r'}}$$

$$\leq c \, \left\|u|W_p^{n/p}\right\|^{\frac{p}{r}} \, |\Omega|^{\frac{1}{r'}} \left\|u|W_p^{n/p}\right\|^{\frac{p'}{r'}}$$

$$\leq c \, |\Omega|^{\frac{r-p}{p'-p}} \left\|u|W_p^{n/p}\right\|^r ,$$

such that finally,

$$\int\limits_{\Omega} |u(x)|^r dx \leq c \, |\Omega|^{\min(\frac{r-p}{p'-p},1)} \left(\sum_{|\alpha| \leq \frac{n}{p}} \int\limits_{\Omega} |D^\alpha u(x)|^p dx \right)^{\frac{r}{p}} \qquad (2.52)$$

for all r, $p \leq r < \infty$. In order to remove the dependence on Ω, suppose that we cover $\mathbb{R}^n$ with cubes Q_m, $m \in \mathbb{Z}^n$, of side length $b > 1$, centered at $m \in \mathbb{Z}^n$, and with sides parallel to the axes of coordinates. Thus $|Q_m| = b^n$ for all $m \in \mathbb{Z}^n$, and we have a controlled overlapping, $\#\{l \in \mathbb{Z}^n \ : Q_l \cap Q_m \neq \emptyset\} \leq c$, $m \in \mathbb{Z}^n$; for a finite set M the number of its elements is denoted by $\# M$, as usual. Let dQ, $d > 0$, denote the cube with the same centre as Q and side length db. Then, with a suitably adapted partition of unity, (2.52) with $\Omega = Q_m$ and $r \geq p$ finally complete the proof of (ii),

$$\|u|L_r\|^r = \int\limits_{\mathbb{R}^n} |u(x)|^r dx = \sum_{m \in \mathbb{Z}^n} \int\limits_{b^{-1}Q_m} |u(x)|^r dx$$

$$\leq c \sum_{m \in \mathbb{Z}^n} |Q_m|^{\min(\frac{r-p}{p'-p},1)} \left(\sum_{|\alpha| \leq \frac{n}{p}} \int\limits_{Q_m} |D^\alpha u(x)|^p dx \right)^{\frac{r}{p}}$$

$$\leq c' \, b^{n \min(\frac{r-p}{p'-p},1)} \left(\sum_{|\alpha| \leq \frac{n}{p}} \sum_{m \in \mathbb{Z}^n} \int\limits_{Q_m} |D^\alpha u(x)|^p dx \right)^{\frac{r}{p}}$$

$$\leq c'' \left(\sum_{|\alpha| \leq \frac{n}{p}} \int\limits_{\mathbb{R}^n} |D^\alpha u(x)|^p dx \right)^{\frac{r}{p}} = c'' \, \left\|u|W_p^{n/p}\right\|^r .$$

Step 4. We prove (2.38). Note that this immediately implies (2.39) as well as (2.40) for $r = \infty$,

$$\|u|L_\infty\| \leq c \, \|u|W_p^k\| , \qquad (2.53)$$

and for $p \leq r < \infty$,

$$\|u|L_r\|^r = \int_{\mathbb{R}^n} |u(x)|^p \, |u(x)|^{r-p} \, dx \; \leq \; \|u|C\|^{r-p} \, \|u|L_p\|^p$$

$$\leq c \, \|u|W_p^k\|^{r-p} \, \|u|W_p^k\|^p \; = \; c \, \|u|W_p^k\|^r.$$

In view of the above argument it is sufficient to deal with the case of $u \in C_0^\infty \cap W_p^k$ such that supp $u \subset \Omega$ with $|\Omega| < \infty$, and to check the dependence of the constants upon Ω. For simplicity we may even assume from the beginning that Ω is the above cube with edges parallel to the axes of coordinates, and side-length $b \geq 1$. Let $x \in \Omega$ and use the Taylor expansion for u in Ω,

$$u(y) = \sum_{|\alpha| \leq k-1} c_\alpha D^\alpha u(x)(y-x)^\alpha + R_x(y), \quad y \in \Omega,$$

where we may choose the integral representation for the remainder term,

$$R_x(y) = \sum_{|\alpha|=k} c_{\alpha,k}(y-x)^\alpha \int_0^1 (1-t)^{k-1} D^\alpha u\left((1-t)x + ty\right) dt.$$

Thus, we have for any $y \in \Omega$,

$$\int_\Omega |R_x(y)| \, dx \leq \sum_{|\alpha|=k} c_{\alpha,k} \int_\Omega |y-x|^k \int_0^1 (1-t)^{k-1} |D^\alpha u\left((1-t)x+ty\right)| \, dt \, dx$$

$$\leq c \sum_{|\alpha|=k} \int_0^1 (1-t)^{-n-1} \int_{\Omega_t} |y-z|^k |D^\alpha u(z)| \, dz \, dt,$$

where $\Omega_t = \{z \in \mathbb{R}^n : z = (1-t)x + ty, \ x \in \Omega\}$. Our special choice of Ω then allows us to estimate further,

$$\int_\Omega |R_x(y)| \, dx \leq c \sum_{|\alpha|=k} \int_\Omega |y-z|^k \, |D^\alpha u(z)| \int_0^{1-\frac{|z-y|}{b}} (1-t)^{-n-1} \, dt \, dz$$

$$\leq c' \, b^n \sum_{|\alpha|=k} \int_\Omega |y-z|^{k-n} \, |D^\alpha u(z)| \, dz. \tag{2.54}$$

Assume $k = n$, $p = 1$, then (2.54) leads to

$$\int_\Omega |R_x(y)| dx \leq c \, b^n \sum_{|\alpha|=k} \|D^\alpha u|L_1\| \; \leq \; c' \, b^n \, \|u|W_1^n\|, \tag{2.55}$$

whereas for $k > \frac{n}{p}$, $p \geq 1$, we conclude by Hölder's inequality for any $y \in \Omega$,

$$
\int_\Omega |R_x(y)| dx \leq c\, b^n \sum_{|\alpha|=k} \|D^\alpha u | L_p\| \left(\int_\Omega |y-z|^{(k-n)p'} dz \right)^{1/p'}
$$

$$
\leq c'\, b^n \|u|W_p^k\| \left(\int_0^{2b} \varrho^{n-1+(k-n)p'} d\varrho \right)^{1/p'}
$$

$$
\leq c''\, b^{n+k-\frac{n}{p}} \|u|W_p^k\|, \tag{2.56}
$$

(suitably adapted for $p=1$). The last integral converges, since $k > \frac{n}{p}$ implies $n+(k-n)p' > 0$. Concerning the Taylor polynomial we estimate similarly,

$$
\int_\Omega \Big| \sum_{|\alpha|\leq k-1} c_\alpha D^\alpha u(x)(y-x)^\alpha \Big| dx \leq c\, b^{k-1} \sum_{|\alpha|\leq k-1} \int_\Omega |D^\alpha u(x)| dx
$$

$$
\leq c\, b^{k-1+\frac{n}{p'}} \sum_{|\alpha|\leq k-1} \|D^\alpha u | L_p\|
$$

$$
\leq c'\, b^{n-1+k-\frac{n}{p}} \|u|W_p^k\|, \tag{2.57}
$$

for any $y \in \Omega$. Finally, since

$$
b^n\, |u(y)| = \int_\Omega |u(y)| dx \leq \int_\Omega \Big(\Big| \sum_{|\alpha|\leq k-1} c_\alpha D^\alpha u(x)(y-x)^\alpha \Big| + |R_x(y)| \Big) dx,
$$

estimates (2.55), (2.56) and (2.57) yield

$$
|u(y)| \leq c\, b^{k-\frac{n}{p}} \|u|W_p^k\|
$$

for any $y \in \Omega$, where either $k > \frac{n}{p}$, $p \geq 1$, or $k = n$, $p = 1$. This yields (2.38) first for $u \in C_0^\infty \cap W_p^k$ with supp $u \subset \Omega$; completely parallel to the end of Step 3 this can then be extended to the general situation.

Step 5.   It remains to prove

$$
W_p^k \hookrightarrow \mathrm{Lip}^a \quad \text{if} \quad \left\{ \begin{array}{ll} 0 < a \leq k-\frac{n}{p}, & \frac{n}{p} < k < \frac{n}{p}+1 \\ 0 < a < 1 & , \; k = \frac{n}{p}+1 \\ 0 < a \leq 1 & , \; p=1, \; k=n+1 \end{array} \right\}. \tag{2.58}
$$

In view of (2.38) and (2.25) it is sufficient to verify

$$
\sup_{\substack{x,y \in \mathbb{R}^n \\ 0 < |x-y| < 1}} \frac{|u(x)-u(y)|}{|x-y|^a} \leq c\, \|u|W_p^k\|
$$

in the cases explicated in (2.58). Moreover, we claim that it is sufficient to deal with the case $k = 1$, that is, to show

$$\sup_{0<|x-y|<1} \frac{|u(x) - u(y)|}{|x-y|^a} \leq c \sum_{|\alpha|\leq 1} \|D^\alpha u | L_r\|, \tag{2.59}$$

where $n < r \leq \infty$, $0 < a \leq 1 - \frac{n}{r}$. Taking (2.59) for granted, we can choose r such that $k - \frac{n}{p} = 1 - \frac{n}{r}$ when $\frac{n}{p} < k < \frac{n}{p} + 1$, and (2.45) applied to $D^\beta u$, $|\beta| = 1$, implies the first case in (2.58). When $k = \frac{n}{p} + 1$, we choose r such that $p < r < \infty$ and $0 < 1 - \frac{n}{r} < 1$; hence (2.44) completes the corresponding argument. Finally, dealing with the third case in (2.58), that is, $p = 1$, $k = n + 1$, we choose $r = \infty$ and use (2.40) together with (2.59). Consequently, it is sufficient to prove (2.59). We deal with the local setting first and consider a unit cube Ω with edges parallel to the axes of coordinates. Let $x, y \in \Omega$, with $0 < |x - y| = \delta < 1$. We denote by Ω_δ the cube with faces parallel to those of Ω, side length δ, and such that $x, y \in \overline{\Omega_\delta} \subset \Omega$. Let $\xi \in \Omega_\delta$, then

$$u(x) = u(\xi) - \int_0^1 \frac{d}{dt} u(x + t(\xi - x)) dt,$$

and

$$|u(x) - u(\xi)| \leq \sqrt{n}\, \delta \int_0^1 |\nabla u(x + t(\xi - x))|\, dt,$$

as $x, \xi \in \overline{\Omega_\delta}$. Straightforward calculation then gives,

$$\left| u(x) - \delta^{-n} \int_{\Omega_\delta} u(\xi) d\xi \right| = \delta^{-n} \left| \int_{\Omega_\delta} (u(x) - u(\xi)) d\xi \right|$$

$$\leq \delta^{-n} \int_{\Omega_\delta} |u(x) - u(\xi)|\, d\xi$$

$$\leq \sqrt{n}\, \delta^{1-n} \int_{\Omega_\delta} \int_0^1 |\nabla u(x + t(\xi - x))|\, dt\, d\xi$$

$$\leq \sqrt{n}\, \delta^{1-n} \int_0^1 t^{-n} \int_{\Omega_{\delta,t}} |\nabla u(\eta)|\, d\eta\, dt$$

$$\leq \sqrt{n}\, \delta^{1-n} \|\nabla u | L_r\| \int_0^1 t^{-n} |\Omega_{\delta,t}|^{\frac{1}{r'}} dt$$

$$\leq c\, \delta^{1-\frac{n}{r}} \sum_{|\alpha|\leq 1} \|D^\alpha u | L_r\| \int_0^1 t^{-\frac{n}{r}} dt$$

$$\leq c' \, \delta^{1-\frac{n}{r}} \sum_{|\alpha|\leq 1} \|D^\alpha u|L_r\|,$$

where we applied Hölder's inequality for $r < \infty$ and used the assumption $r > n$; the modification for $r = \infty$ is obvious. Likewise we have

$$\left| u(y) - \delta^{-n} \int_{\Omega_\delta} u(\xi)\mathrm{d}\xi \right| \leq c \, \delta^{1-\frac{n}{r}} \sum_{|\alpha|\leq 1} \|D^\alpha u|L_r\|,$$

such that the triangle inequality together with the substitution $\delta = |x - y|$ leads to

$$|u(x) - u(y)| \leq c \, |x - y|^{1-\frac{n}{r}} \sum_{|\alpha|\leq 1} \|D^\alpha u|L_r\|,$$

for all $x, y \in \Omega$, $x \neq y$. We consider a covering of $\mathbb{R}^n$ with cubes Q_m as described above and centered at $m \in \mathbb{Z}^n$, then for arbitrary $x, y \in \mathbb{R}^n$ with $|x - y| < \frac{b}{2}$ there is at least one cube Q_m such that $x, y \in Q_m$ and we can apply the above argument,

$$\sup_{\substack{x,y\,\in\,\mathbb{R}^n \\ 0 < |x-y| < 1}} \frac{|u(x) - u(y)|}{|x-y|^{1-\frac{n}{r}}}$$

$$\leq \sup_{m\in\mathbb{Z}^n} \sup_{\substack{x,y\,\in\,Q_m \\ 0 < |x-y| < \frac{b}{2}}} \frac{|u(x) - u(y)|}{|x-y|^{1-\frac{n}{r}}} + c \sup_{x\in\mathbb{R}^n} |u(x)|$$

$$\leq c' \sum_{|\alpha|\leq 1} \|D^\alpha u|L_r\| + c \, \|u|C\|$$

$$\leq c'' \sum_{|\alpha|\leq 1} \|D^\alpha u|L_r\|$$

using (2.38) with $r > n$ in the final step. This proves (2.59) and finishes the whole proof. $\qquad\square$

Remark 2.29 As already mentioned in the introduction, a lot of work has been done since Sobolev's pioneering paper [Sob38], dealing with extensions, sharpness, best constants, dependence upon the (geometry of the) underlying domain of the above embeddings or inequalities, respectively. We do not want to repeat this here, but refer to the introduction.

The result of Brézis-Wainger [BW80], mentioned in the introduction (1.14) obviously complements (2.42). With the notation introduced in Definition 2.26 and taking into consideration that $W_p^m = H_p^m$, $m \in \mathbb{N}_0$, $1 < p < \infty$, (1.14) can be reformulated as

$$W_p^{1+n/p} \hookrightarrow \mathrm{Lip}^{(1,-\frac{1}{p'})}, \quad 1 < p < \infty, \quad \frac{1}{p} + \frac{1}{p'} = 1. \tag{2.60}$$

Example 2.30 Concerning the sharpness of Theorem 2.28 we restrict our-selves to an example merely, since the main idea of envelopes just relies on those cases where we do not have the corresponding embeddings. We explain this in detail in the next chapters. However, this immediately proves the sharpness of the above assertions, too (and is also known already).

Dealing with the critical case (ii) when $k = \frac{n}{p} = 1$ and $n > 1$, see (2.44) and (2.38), we present a family of functions h_ν that belong to W_n^1, but not to L_∞, as long as $0 < \nu < 1 - \frac{1}{n}$. For $x \in \mathbb{R}^n$, consider the radial functions

$$
h_\nu(x) = \begin{cases} \log^\nu \left(1 + \dfrac{1}{|x|}\right) - \log^\nu 2, & 0 < |x| < 1 \\ 0 & , \quad \text{otherwise} \end{cases}, \quad x \in \mathbb{R}^n. \qquad (2.61)
$$

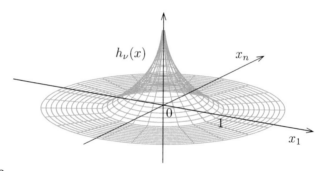

Figure 9

Plainly, $h_\nu \notin L_\infty$ whenever $\nu > 0$, whereas for all $0 < r < \infty$,

$$
\|h_\nu | L_r\|^r = \int_{\mathbb{R}^n} |h_\nu(x)|^r \, dx
$$

$$
\leq c_1 + c_2 \int_0^1 \varrho^{n-1} \log^{r\nu} \left(1 + \frac{1}{\varrho}\right) d\varrho
$$

$$
\leq c_1 + c_3 \int_0^1 \varrho^{n-1-\varepsilon} d\varrho \leq c_4 < \infty
$$

for all $n \in \mathbb{N}$ and sufficiently small $\varepsilon > 0$. Using the abbreviation D_k intro-duced in Step 1 of the proof of Theorem 2.28, straightforward calculations lead to the estimates

$$\|D_k h_\nu | L_n\|^n = \int_{\mathbb{R}^n} |D_k h_\nu(x)|^n \, dx$$

$$\leq c_1 \int_0^1 \varrho^{n-1} \frac{\log^{n(\nu-1)}\left(1 + \frac{1}{\varrho}\right)}{(\varrho^2 + \varrho)^n} \, d\varrho$$

$$\leq c_2 \int_0^1 \left(\frac{\varrho}{\varrho^2 + \varrho}\right)^{n-1} \frac{\log^{n(\nu-1)}\left(1 + \frac{1}{\varrho}\right)}{\varrho^2 + \varrho} \, d\varrho$$

$$\leq c_2 \int_0^1 \frac{\log^{n(\nu-1)}\left(1 + \frac{1}{\varrho}\right)}{\varrho^2 + \varrho} \, d\varrho$$

$$\leq c_3 \left. \log^{n(\nu-1)+1}\left(1 + \frac{1}{\varrho}\right) \right|_0^1 \leq c_4 < \infty$$

assuming that $\nu < 1 - \frac{1}{n}$. Hence,

$$\|h_\nu | W_n^1\| = \left(\sum_{|\alpha| \leq 1} \|D^\alpha h_\nu | L_n\|^n\right)^{1/n}$$

$$\leq c \, \|h_\nu | L_n\| + c \sum_{k=1}^n \|D_k h_\nu | L_n\| \leq c' < \infty$$

for $0 < \nu < 1 - \frac{1}{n}$.

Another well-known example following the same idea, i.e., $g \in W_n^1 \setminus L_\infty$, is given by

$$g(x) = \begin{cases} \log\left(\log\left(1 + \frac{1}{|x|}\right)\right) - \log\log 2, & 0 < |x| < 1 \\ 0 & , \quad \text{otherwise} \end{cases}, \quad x \in \mathbb{R}^n.$$

Chapter 3

The growth envelope function $\mathcal{E}_\mathsf{G}$

We already mentioned in our introductory remarks that characterisations like (1.4) gave reason to study the behaviour of the non-increasing rearrangement f^* of a function f, in particular, when these spaces contain essentially unbounded functions. This leads to the concept of growth envelopes. Our results for spaces of type $B_{p,q}^s$ or $F_{p,q}^s$ are postponed to Part II; we start with some simple features and examples in order to give a better feeling for what is really "measured" by growth envelopes. For that reason we test our new envelope tool on rather classical spaces like Lorentz (-Zygmund) spaces first – before arriving at more surprising results in Part II. Moreover, there is also an interesting point at the end of this chapter: the recognition of growth envelope functions in terms of fundamental functions in rearrangement-invariant spaces. Finally, preparing some later discussion of global versus local behaviour in Section 10.3 we already add some "higher-order" and "weighted" examples in Section 3.4.

3.1 Definition and basic properties

Definition 3.1 *Let $[\mathcal{R}, \mu]$ be a measure space and X a quasi-normed function space on $\mathcal{R}$. The* growth envelope function *$\mathcal{E}_\mathsf{G}^X : (0, \infty) \to [0, \infty]$ of X is defined by*

$$\mathcal{E}_\mathsf{G}^X(t) := \sup_{\|f|X\| \le 1} f^*(t), \quad t > 0. \tag{3.1}$$

We adopt the usual convention to put $\mathcal{E}_\mathsf{G}^X(\tau) := \infty$ when $\{f^*(\tau) : \|f|X\| \le 1\}$ is not bounded from above for some $\tau > 0$.

Remark 3.2 Note that (3.1) immediately causes some problem when taking into account that we shall always deal with equivalent (quasi-) norms in the underlying function space (rather than a fixed one). Assume we have two different, but equivalent (quasi-) norms $\| \cdot |X\|_1$ and $\| \cdot |X\|_2$ in X. Then every function $f \in X$ with $\|f|X\|_1 \le 1$, $f \not\equiv 0$, is connected with some $g_f := cf$, where $c = \|f|X\|_1 / \|f|X\|_2$, $\|g_f|X\|_2 \le 1$, and $g_f^* = cf^*$, leading

39

to a different, but equivalent expression for $\mathcal{E}_G^X$. So, strictly speaking, we are concerned with equivalence classes of growth envelope functions, where we choose one representative

$$\mathcal{E}_G^X(t) \sim \sup_{\|f|X\| \leq 1} f^*(t), \quad t > 0.$$

However, we shall not make this difference between equivalence class and representative in the sequel. Furthermore, by (3.1) the growth envelope $\mathcal{E}_G^X(t)$ is defined for all values $t > 0$, but at the moment we are only interested in local characterisations (singularities) of the spaces referring to small values of $t > 0$, say, $0 < t < 1$, whereas questions of global behaviour $(t \to \infty)$ are postponed to Section 10.3. This preference for local studies also implies that we could formally transfer many of our results from spaces on $\mathbb{R}^n$ to their counterparts on bounded domains, or, more precisely, from measure spaces $[\mathcal{R}, \mu]$ with $\mu(\mathcal{R}) = \infty$ to finite measure spaces. The necessary modifications in the case of our examples in Section 3.2 are obvious; we shall thus mainly deal with function spaces on $\mathbb{R}^n$ in the sequel.

We briefly discuss the obvious question whether the growth envelope function $\mathcal{E}_G^X$ is always finite for $t > 0$ or what necessary/sufficient conditions on X (or the underlying measure space) imply this; recall notation (2.2).

Lemma 3.3 *Let $[\mathcal{R}, \mu]$ be a measure space.*

(i)   *There are function spaces X on $\mathcal{R}$ which do not have a growth envelope function in the sense that $\mathcal{E}_G^X(t)$ is not finite for $t > 0$.*

(ii)   *Let X be a (quasi-) normed function space on $\mathcal{R}$. Then $\mathcal{E}_G^X(t)$ is finite for any $t > 0$ if, and only if,*

$$\sup_{\|f|X\| \leq 1} \mu_f(\lambda) \longrightarrow 0 \qquad \text{for} \quad \lambda \to \infty. \tag{3.2}$$

P r o o f :   Concerning (i), we take a simple counter-example, obviously such that $X \not\hookrightarrow L_\infty$. Let $[\mathcal{R}, \mu] = [\mathbb{R}^n, |\cdot|]$ and put $X := L_\infty(\langle x \rangle^{-1})$. Hence $f(x) = \langle x \rangle$ belongs to $X = L_\infty(\langle x \rangle^{-1})$, $\|f|X\| = 1$, but $f^*(t)$ is not finite for any $t > 0$.
As for (ii), we first prove that $\mathcal{E}_G^X(t) < \infty$ for any $t > 0$ implies (3.2). Hence we have for any $t > 0$ that there is some $M_t > 0$ such that for all $f \in X$, $\|f|X\| \leq 1$, we have $f^*(t) \leq M_t$. Thus there is for any $t > 0$ some $M_t > 0$ such that for all $f \in X$, $\|f|X\| \leq 1$, we know $\mu_f(\lambda) \leq t$ for any $\lambda > M_t$. In other words, for any $t > 0$ there is some $M_t > 0$ such that for all $\lambda > M_t$ the expression $\sup_{\|f|X\| \leq 1} \mu_f(\lambda)$ is bounded from above by t. But this is nothing else than a reformulation of (3.2). The converse can be shown by the same (standard) argument.   □

Hence the definition of $\mathcal{E}_\mathsf{G}^X$ is non-trivial and reasonable. We now collect a few elementary properties of it. Simplifying technical matters in the sequel we introduce the number τ_0 by

$$\tau_0 = \tau_0^\mathsf{G}(X) := \sup\left\{t > 0 \;:\; \mathcal{E}_\mathsf{G}^X(t) > 0\right\}. \tag{3.3}$$

Note that $\mathcal{E}_\mathsf{G}^X(t) = 0$ for some $t > 0$ implies $f^*(t) = 0$ for all $f \in X$, $\|f|X\| \leq 1$; hence – by some scaling argument – $g^*(t) = 0$ for all $g \in X$. But then X contains only functions having a support with finite measure, i.e., $\mu\left(\{x \in \mathcal{R} : |g(x)| > 0\}\right) \leq t$ for all $g \in X$. This is true, in particular, when $\mu(\mathcal{R}) \leq t$.

Proposition 3.4 *Let* $[\mathcal{R}, \mu]$ *be a measure space and* X *a quasi-normed function space on* $\mathcal{R}$.

(i) $\mathcal{E}_\mathsf{G}^X$ *is monotonically decreasing and right-continuous,* $\mathcal{E}_\mathsf{G}^X = \left(\mathcal{E}_\mathsf{G}^X\right)^*$.

(ii) *If* $\mathcal{R}$ *has finite measure, i.e.,* $\mu(\mathcal{R}) < \infty$, *then* $\mathcal{E}_\mathsf{G}^X(t) = 0$ *for* $t > \mu(\mathcal{R})$ *and any function space* X *on* $\mathcal{R}$.

(iii) *We have* $X \hookrightarrow L_\infty$ *if, and only if,* $\mathcal{E}_\mathsf{G}^X(\cdot)$ *is bounded, i.e.,* $\sup\limits_{t>0} \mathcal{E}_\mathsf{G}^X(t) = \lim\limits_{t\downarrow 0} \mathcal{E}_\mathsf{G}^X(t)$ *is finite. In that case,*

$$\mathcal{E}_\mathsf{G}^X(0) := \lim_{t\downarrow 0} \mathcal{E}_\mathsf{G}^X(t) = \|\mathrm{id} : X \to L_\infty\|.$$

(iv) *Let* $X_i = X_i(\mathcal{R})$, $i = 1, 2$, *be function spaces on* $\mathcal{R}$. *Then* $X_1 \hookrightarrow X_2$ *implies that there is some positive constant* c *such that for all* $t > 0$,

$$\mathcal{E}_\mathsf{G}^{X_1}(t) \;\leq\; c\,\mathcal{E}_\mathsf{G}^{X_2}(t).$$

One may choose $c = \|\mathrm{id} : X_1 \to X_2\|$ *in that case.*

(v)   *Let* $\varkappa : (0, \infty) \to [0, \infty)$ *be a non-negative function and* $\mathcal{E}_\mathsf{G}^X(t) < \infty$ *for* $t > 0$. *Then* $\varkappa(\cdot)$ *is bounded on* $(0, \tau_0)$ *if, and only if, there exists* $c > 0$ *such that for all* $f \in X$, $\|f|X\| \leq 1$,

$$\sup_{0 < t < \tau_0} \frac{\varkappa(t)}{\mathcal{E}_\mathsf{G}^X(t)}\, f^*(t) \;\leq\; c. \tag{3.4}$$

P r o o f :   As for (i), note that f^* is monotonically decreasing for any $f \in X$; thus also $\mathcal{E}_\mathsf{G}^X$ is monotonically decreasing by standard arguments. This implies $\left(\mathcal{E}_\mathsf{G}^X\right)^* = \mathcal{E}_\mathsf{G}^X$, too. Likewise the right-continuity of any f^* implies that of $\mathcal{E}_\mathsf{G}^X$. Assertion (ii) is clear, because obviously $f^*(t) = 0$ for any $t > \mu(\mathcal{R})$ and any $f \in X$ in that case.

We prove (iii). Assume first that $X \hookrightarrow L_\infty$, i.e., there exists $c > 0$ such that for all $f \in X$, $\|f|L_\infty\| \le c \|f|X\|$. Then $f^*(0) = \lim\limits_{t \downarrow 0} f^*(t) = \|f|L_\infty\| < \infty$, which implies

$$\mathcal{E}_G^X(0) := \lim_{t \downarrow 0} \mathcal{E}_G^X(t) = \sup_{t > 0} \sup_{\|f|X\| \le 1} f^*(t) = \sup_{\|f|X\| \le 1} f^*(0) = \sup_{\|f|X\| \le 1} \|f|L_\infty\|$$

$$\le c \sup_{\|f|X\| \le 1} \|f|X\| = c.$$

Conversely, let $X \not\hookrightarrow L_\infty$. Then by [BS88, Ch. 1, Thm. 1.8] $X \not\subset L_\infty$ for any Banach function space X, and there is nothing else to prove. Otherwise, when $X \subset L_\infty$, but $\mathrm{id} : X \to L_\infty$ is not continuous, one finds a sequence $(t_n)_n$ such that $\mathcal{E}_G^X(t_n) > n$, $n \in \mathbb{N}$, and the monotonicity of $\mathcal{E}_G^X$ leads to $\sup\limits_{t > 0} \mathcal{E}_G^X(t) = \infty$.

Verifying (iv), let $f \in X_1$, $\|f|X_1\| \le 1$, and put $c := \|\mathrm{id} : X_1 \to X_2\|$. Then $(\frac{1}{c}f) \in X_2$, $\|(\frac{1}{c}f)|X_2\| \le 1$, and $\frac{1}{c}f^*(t) = (\frac{1}{c}f)^*(t) \le \mathcal{E}_G^{X_2}(t)$. Consequently for any $f \in X_1$, $\|f|X_1\| \le 1$, we obtain $f^*(t) \le c \, \mathcal{E}_G^{X_2}(t)$, implying (iv).

We first prove the necessity of (3.4) for the boundedness of $\varkappa$. Thus let $\varkappa$ be a positive and bounded function, i.e., there is some $c > 0$ such that for all $0 < t < \tau_0$ we have $\varkappa(t) \le c$. On the other hand, by definition $\mathcal{E}_G^X(t) \ge f^*(t)$ for any $t > 0$ and any $f \in X$, $\|f|X\| \le 1$; hence there is some $c > 0$ such that for all $f \in X$, $\|f|X\| \le 1$, and all $0 < t < \tau_0$,

$$\frac{\varkappa(t)}{\mathcal{E}_G^X(t)} \, f^*(t) \le c.$$

This implies (3.4). It remains to show the sufficiency of (3.4) for the boundedness of $\varkappa$. But obviously (3.4) leads to the existence of some $c > 0$ such that

$$\sup_{\|f|X\| \le 1} \sup_{0 < t < \tau_0} \frac{\varkappa(t)}{\mathcal{E}_G^X(t)} \, f^*(t) \le c.$$

We may rewrite this as

$$\sup_{0 < t < \tau_0} \frac{\varkappa(t)}{\mathcal{E}_G^X(t)} \sup_{\|f|X\| \le 1} f^*(t) \le c,$$

but by the definition of $\mathcal{E}_G^X$ this reduces to $\sup\limits_{0 < t < \tau_0} \varkappa(t) \le c$, i.e., $\varkappa$ is bounded. $\qquad\square$

Remark 3.5 We shall see in the next section that some counterpart of (iv) in the sense of (iii), i.e., that some relation of the envelope functions implies some (continuous) embedding for the corresponding spaces, cannot hold in general; see Remark 3.13.

Concerning (v), one may prove even more, namely that in some sense $\mathcal{E}_G^X$ is the only such function with the property described above.

Corollary 3.6 *Let* $[\mathcal{R}, \mu]$ *be a measure space,* X *a (quasi-) normed function space over* $\mathcal{R}$, *and* $\psi : (0, \infty) \to (0, \infty]$ *a positive, monotonically decreasing function with the following property: For any non-negative function* $\varkappa : (0, \infty) \to [0, \infty)$, $\varkappa(\cdot)$ *is bounded on* $(0, \tau_0)$ *if, and only if, there exists* $c > 0$ *such that for all* $f \in X$, $\|f|X\| \leq 1$,

$$\sup_{0 < t < \tau_0} \frac{\varkappa(t)}{\psi(t)} f^*(t) \leq c. \tag{3.5}$$

Then $\psi \sim \mathcal{E}_G^X$, *i.e., there are numbers* $c_2 > c_1 > 0$ *such that whenever* $0 < t < \tau_0$,

$$c_1 \psi(t) \leq \mathcal{E}_G^X(t) \leq c_2 \psi(t). \tag{3.6}$$

P r o o f : We have by (3.5) with $\varkappa \equiv 1$ that there is some $c > 0$ such that for all $f \in X$, $\|f|X\| \leq 1$,

$$\sup_{0 < t < \tau_0} \frac{1}{\psi(t)} f^*(t) \leq c,$$

i.e., that there exists $c > 0$ such that for all $t \in (0, \tau_0)$ and all $f \in X$, $\|f|X\| \leq 1$, $f^*(t) \leq c\,\psi(t)$. Hence by the definition of $\mathcal{E}_G^X(t)$, there exists $c > 0$ such that for all $0 < t < \tau_0$,

$$\mathcal{E}_G^X(t) \leq c\,\psi(t). \tag{3.7}$$

It remains to show the converse. Let $\varkappa$ be a non-negative function. Then we obtain by (3.4) that $\varkappa$ is bounded if, and only if, there is some $c > 0$ such that for all $f \in X$, $\|f|X\| \leq 1$,

$$\sup_{0 < t < \tau_0} \frac{\varkappa(t)}{\mathcal{E}_G^X(t)} f^*(t) \leq c, \qquad \text{i.e.,} \qquad \sup_{0 < t < \tau_0} \frac{\varkappa(t)}{\psi(t)} \frac{\psi(t)}{\mathcal{E}_G^X(t)} f^*(t) \leq c.$$

Now we conclude from (3.5) with $\tilde{\varkappa} := \varkappa \psi / \mathcal{E}_G^X$ that this happens if, and only if, $\tilde{\varkappa}$ is bounded. Hence we know that $\varkappa$ is bounded if, and only if, $\tilde{\varkappa} = \varkappa \psi / \mathcal{E}_G^X$ is bounded. Let, in particular, $\varkappa \equiv 1$; then this provides the inequality converse to (3.7) and ends the proof. $\qquad\square$

Remark 3.7 Triebel discussed earlier the closely related concept of a *growth function* ψ, where $[\mathcal{R}, \mu] = [\mathbb{R}^n, \ell_n]$ and $X = B_{p,q}^s$ or $X = F_{p,q}^s$. In our notation this means nothing else than $\psi \sim 1/\mathcal{E}_G^X$.

Later on we shall work with measure spaces $[\mathcal{R}, \mu]$ and function spaces X such that there is a representative in the equivalence class of $\mathcal{E}_G^X$ which is continuous near 0. In particular, for our purpose it was sufficient to obtain

$$\mathcal{E}_G^X \left(2^{-j}\right) \sim \mathcal{E}_G^X \left(2^{-j+1}\right) \tag{3.8}$$

for some $j_0 \in \mathbb{N}$ and all $j \geq j_0$. In the case of $[\mathcal{R}, \mu] = [\mathbb{R}^n, \ell_n]$ we can, for instance, assume that X additionally satisfies

$$\left\| f \left(2^{-\frac{1}{n}} \cdot\right) | X \right\| \leq c \, \|f|X\| \tag{3.9}$$

for some $c > 0$ and all $f \in X$; this generalises [Tri01, (12.38)]. The monotonicity of $\mathcal{E}_G^X$, see Proposition 3.4(i), immediately yields "$\geq$" in (3.8), whereas the converse inequality uses functions $f_n(x) := f(2^{-\frac{1}{n}} x)$ built upon $f \in X$, say, with $\|f|X\| \leq 1$. Plainly $f_n^*(2t) = f^*(t)$, the rest is covered by (3.9). Of course, any number $\lambda_0 < 1$ replacing $2^{-\frac{1}{n}}$ in (3.9) would do, modifying (3.8) appropriately.

Lemma 3.8 *Let the measure space $[\mathcal{R}, \mu]$ and the function space X be such that for any $f \in X$ there is some $f_\mu \in X$ with $f_\mu^*(2^{-j+1}) = f^*(2^{-j})$ for some $j_0 \in \mathbb{N}_0$ and all $j \geq j_0$, and $\|f_\mu|X\| \leq c \, \|f|X\|$. Then (3.8) holds for all $j \geq j_0$,*

$$\mathcal{E}_G^X \left(2^{-j}\right) \sim \mathcal{E}_G^X \left(2^{-j+1}\right).$$

Proof: The proof copies the idea of $[\mathbb{R}^n, \ell_n]$ with $f_\mu = f_n$ described above. Proposition 3.4(i) implies "$\geq$" in (3.8), whereas we conclude from

$$\mathcal{E}_G^X \left(2^{-j}\right) = \sup_{\|f|X\| \leq 1} f^* \left(2^{-j}\right) \leq c \sup_{\|f_\mu|X\| \leq 1} f_\mu^* \left(2^{-j+1}\right) \leq c \, \mathcal{E}_G^X \left(2^{-j+1}\right)$$

the converse estimate. □

Remark 3.9 We shall need the above assertion only in connection with the continuity of (a function equivalent to) $\mathcal{E}_G^X$ near 0. Assume, for instance, that the measure μ has the property that $\mu(\tau A) = h_\mu(\tau) \, \mu(A)$ for small τ and A in the σ-algebra of $\mathcal{R}$. Then we can put f_μ such that $f_\mu(x) := f(\tau_0 x)$, if τ_0 satisfies $h_\mu(\tau_0) = \frac{1}{2}$. (In case of $[\mathcal{R}, \mu] = [\mathbb{R}^n, \ell_n]$ this refers to $h_\mu(\tau) = \tau^n$ and thus $\tau_0 = 2^{-\frac{1}{n}}$.) Consequently we had to pose the restriction on X that $\|f(\tau_0 \cdot)|X\| \leq c \, \|f|X\|$ for all $f \in X$, see (3.9).

3.2   Examples: Lorentz spaces

For convenience we adopt the following notation. Here and in the sequel we shall mean by $\mathrm{Im}(\mu) = [0, \mu(\mathcal{R})]$ that the range of μ is the whole interval $[0, \mu(\mathcal{R})]$, i.e., that for every number $s \in [0, \mu(\mathcal{R})]$ there is some $A_s \subset \mathcal{R}$ in the σ-algebra of $\mathcal{R}$ with $\mu(A_s) = s$.

We start with a preparatory lemma as the "extremal" functions below will be often used in the sequence. So it appears convenient to extract this argument from the subsequent considerations. In the sequel, let

$$K_r(x^0) = \{x \in \mathbb{R}^n : |x - x^0| < r\} \tag{3.10}$$

always stand for the open ball with radius $r > 0$ centred at $x^0 \in \mathbb{R}^n$.

Lemma 3.10 *Let* $0 < s < \mu(\mathcal{R})$ *and* $A_s \subset \mathcal{R}$ *with* $\mu(A_s) = s$.
(i)   *Let* $0 < r \leq \infty$, *and*

$$f_s = s^{-\frac{1}{r}} \chi_{A_s} . \tag{3.11}$$

Then

$$f_s^*(t) = s^{-\frac{1}{r}} \chi_{[0,s)}(t), \quad t > 0, \tag{3.12}$$

and for $0 < p, q \leq \infty$,

$$\|f_s | L_{p,q}\| \sim s^{\frac{1}{p} - \frac{1}{r}} . \tag{3.13}$$

(ii)   *Let* $[\mathcal{R}, \mu] = [\mathbb{R}^n, |\cdot|]$, *and* $A_s = K_{cs^{1/n}}(0)$, *where* $c > 0$ *is suitably chosen such that* $\mu(A_s) = |K_{cs^{1/n}}(0)| = s$. *Let for* $0 < r < \infty$, $\varkappa > 0$, $\gamma \in \mathbb{R}$,

$$f_{s,\varkappa}(x) = s^{-\left(\frac{1}{r} - \varkappa\right)} |x|^{-n\varkappa} \left(1 + |\log|x||\right)^{-\gamma} \chi_{A_s}(x), \quad x \in \mathbb{R}^n. \tag{3.14}$$

Then

$$f_{s,\varkappa}^*(t) \sim s^{-\left(\frac{1}{r} - \varkappa\right)} t^{-\varkappa} \left(1 + |\log t|\right)^{-\gamma} \chi_{[0,s)}(t), \quad t > 0, \tag{3.15}$$

and for $0 < p < \infty$, $0 < q \leq \infty$, $a \in \mathbb{R}$, *and* $\varkappa$ *such that* $0 < \varkappa < \frac{1}{p}$,

$$\|f_{s,\varkappa} | L_{p,q}(\log L)_a\| \sim s^{\frac{1}{p} - \frac{1}{r}} \left(1 + |\log s|\right)^{a - \gamma}, \quad s > 0. \tag{3.16}$$

(iii)   *Let* $[\mathcal{R}, \mu] = [\mathbb{R}^n, |\cdot|]$, *and* $A_s = K_{cs^{1/n}}(0)$ *with* $\mu(A_s) = |K_{cs^{1/n}}(0)| = s$ *for appropriately chosen* $c > 0$. *Let for* $0 < s < 1$, $0 < q < \infty$, $a, \varkappa \in \mathbb{R}$, *with* $0 < \varkappa < -\left(a + \frac{1}{q}\right)$,

$$f_{s,\varkappa}(x) = \left(1 + |\log s|\right)^{\varkappa} |\log|x||^{-\left(a + \frac{1}{q} + \varkappa\right)} \chi_{A_s}(x), \quad x \in \mathbb{R}^n; \tag{3.17}$$

then

$$f_{s,\varkappa}^*(t) \sim (1 + |\log s|)^\varkappa \, |\log t|^{-(a+\frac{1}{q}+\varkappa)} \, \chi_{[0,s)}(t), \quad t > 0, \qquad (3.18)$$

and

$$\|f_{s,\varkappa}|L_{\infty,q}(\log L)_a\| \sim 1, \quad 0 < s < 1. \qquad (3.19)$$

(iv) *Assume* $[\mathcal{R}, \mu] = [K_\varrho(0), |\cdot|]$ *with* $|K_\varrho(0)| = 1$, *and* $A_s = K_{cs^{1/n}}(0)$ *with* $\mu(A_s) = |K_{cs^{1/n}}(0)| = s$ *for appropriately chosen* $c > 0$ *and* $0 < s < 1$. *Let for* $a \geq 0$,

$$h_s(x) = |\log |x||^a \, \chi_{A_s}(x), \quad x \in \mathcal{R}; \qquad (3.20)$$

then

$$h_s^*(t) \sim |\log t|^a \, \chi_{[0,s)}(t), \quad t > 0,$$

and

$$\|h_s|L_{\exp,a}\| \sim 1. \qquad (3.21)$$

P r o o f :   The first assertion in (i) is obvious, see Example 2.4; for the second one note that

$$\|f_s|L_{p,q}\| \sim \left(\int_0^s \left[t^{\frac{1}{p}} s^{-\frac{1}{r}} \right]^q \frac{dt}{t} \right)^{\frac{1}{q}} = s^{-\frac{1}{r}} \left(\int_0^s t^{\frac{q}{p}-1} dt \right)^{\frac{1}{q}} \sim s^{\frac{1}{p}-\frac{1}{r}}, \quad s > 0.$$

We turn to (ii). Again, straightforward calculation leads to (3.15), so that we can further conclude by (2.13),

$$\|f_{s,\varkappa}|L_{p,q}(\log L)_a\| \sim s^{-\left(\frac{1}{r}-\varkappa\right)} \left(\int_0^s t^{\left(\frac{1}{p}-\varkappa\right)q-1} (1 + |\log t|)^{(a-\gamma)q} \, dt \right)^{\frac{1}{q}}$$

$$\sim s^{-\left(\frac{1}{r}-\varkappa\right)} s^{\left(\frac{1}{p}-\varkappa-\varepsilon\right)} (1 + |\log s|)^{(a-\gamma)} \left(\int_0^s t^{\varepsilon q-1} dt \right)^{\frac{1}{q}}$$

$$\sim s^{\frac{1}{p}-\frac{1}{r}} (1 + |\log s|)^{(a-\gamma)}, \quad s > 0,$$

for $0 < \varepsilon < \frac{1}{p} - \varkappa$ and with obvious modification for $q = \infty$. Assertion (3.18) follows directly from our assumptions on the parameters, and again by (2.13) for small s,

$$\|f_{s,\varkappa}|L_{\infty,q}(\log L)_a\| \sim (1 + |\log s|)^\varkappa \left(\int_0^s \left[(1 + |\log t|)^a \, |\log t|^{-(a+\frac{1}{q}+\varkappa)} \right]^q \frac{dt}{t} \right)^{\frac{1}{q}}$$

$$\sim (1 + |\log s|)^\varkappa (1 + |\log s|)^{-\varkappa} \sim 1.$$

It remains to check (iv); recall (2.18), (2.19). Observe that

$$\int\limits_{\mathcal{R}} \exp\left(\lambda|h_s(x)|\right)^{1/a} \mu(dx) = \int\limits_{A_s} |x|^{-\lambda^{1/a}} \mu(dx) + \mu(\mathcal{R} \setminus A_s)$$

$$\leq c' \int\limits_0^{cs^{1/n}} r^{n-1-\lambda^{1/a}} dr + 1 - s \leq C$$

for any $\lambda < n^a$, i.e., $h_s \in L_{\exp,a}$. Moreover, $h_s^*(t) \sim |\log t|^a \, \chi_{[0,s)}(t)$ and thus by (2.19), $\|h_s|L_{\exp,a}\| \sim \sup\limits_{0<t<s} (1 - \log t)^{-a} |\log t|^a \sim 1$. $\square$

Remark 3.11 In general, dealing with function spaces X (of distributions) we shall stick with the assumption $X \subset L_1^{\mathrm{loc}}$ as mentioned above. However, in case of Lorentz (-Zygmund) spaces as given by Definition 2.8 we may incorporate parameters $p, q \leq 1$, always keeping in mind this slight abuse of notation.

Proposition 3.12 *Let $[\mathcal{R}, \mu]$ be a σ-finite measure space with $\mathrm{Im}(\mu) = [0, \mu(\mathcal{R})]$ or a finite non-atomic measure space. Then we obtain for $L_{p,q} = L_{p,q}(\mathcal{R})$, $0 < p, q \leq \infty$ (with $q = \infty$ when $p = \infty$),*

$$\mathcal{E}_G^{L_{p,q}}(t) \sim t^{-\frac{1}{p}}, \qquad 0 < t < \mu(\mathcal{R}). \tag{3.22}$$

Proof: Step 1.   We first prove the assertion for $0 < p < \infty$, $0 < q \leq \infty$. Assume $f \in L_{p,q}$, $\|f|L_{p,q}\| \leq 1$, and let $\tau > 0$ be such that $\tau < \mu(\mathcal{R})$; then by (2.12),

$$1 \geq \|f|L_{p,q}\| \geq \left(\int\limits_0^{\tau} \left[t^{\frac{1}{p}} f^*(t) \right]^q \frac{dt}{t} \right)^{1/q} \geq f^*(\tau) \left(\int\limits_0^{\tau} t^{\frac{q}{p}-1} dt \right)^{1/q}$$

$$= f^*(\tau) \left(\frac{p}{q} \tau^{\frac{q}{p}} \right)^{1/q},$$

and this implies $f^*(\tau) \leq c_{pq} \tau^{-\frac{1}{p}}$ for any τ, $0 < \tau < \mu(\mathcal{R})$. Thus $\mathcal{E}_G^{L_{p,q}}(t) \leq C t^{-\frac{1}{p}}$, $0 < p < \infty$, $0 < q \leq \infty$.

Step 2.   We show the converse inequality for $0 < p < \infty$, $0 < q \leq \infty$. Let $0 < s < \mu(\mathcal{R})$ and $A_s \subset \mathcal{R}$ with $\mu(A_s) = s$. The existence of such an A_s is a consequence of our assumption: while using the first alternative (concerning the range of μ) it is obvious, otherwise, since $\mu(\mathcal{R}) < \infty$ and

Definitions 2.11(vi) and 2.12 then imply that $\chi_\mathcal{R}$ belongs to the μ-measurable functions finite μ-a.e., Remark 2.7 with $f = \chi_\mathcal{R}$ yields the existence of such a set A_s for every s, $0 \leq s \leq \mu(\mathcal{R})$. We apply Lemma 3.10(i) with $r = p$; then $f_s \in L_{p,q}$ with $\|f_s|L_{p,q}\| \sim 1$, and

$$\mathcal{E}_\mathsf{G}^{L_{p,q}}(t) \geq \sup_{s>0} f_s^*(t) = \sup_{s>t} f_s^*(t) \sim \sup_{s>t} s^{-\frac{1}{p}} \sim t^{-\frac{1}{p}}, \quad 0 < t < \mu(\mathcal{R}).$$

Step 3.   It remains to verify (3.22) for $p = q = \infty$. Obviously (3.22) with $p = q = \infty$ is to be understood in the sense that there are constants $c_2 > c_1 > 0$ such that for all $t > 0$

$$c_1 \leq \mathcal{E}_\mathsf{G}^{L_\infty}(t) \leq c_2.$$

It is clear by Definition 3.1 and Proposition 3.4(iii), that $\mathcal{E}_\mathsf{G}^{L_\infty}(t) \leq c$. Conversely, we apply (3.11) with $r = \infty$; then $f_s = \chi_{A_s} \in L_\infty$, $s > 0$, $\|f_s|L_\infty\| = 1$, and for $t > 0$, $\mathcal{E}_\mathsf{G}^{L_\infty}(t) \geq \sup_{s>0} f_s^*(t) = \sup_{s>t} 1 = 1.$ □

Remark 3.13 We return to Remark 3.5. One can easily calculate that for, say, $0 < p, q < \infty$,

$$\mathcal{E}_\mathsf{G}^{L_{p,q}}(t) = \left(\frac{q}{p}\right)^{\frac{1}{q}} t^{-\frac{1}{p}}$$

with fixed $\|\cdot|L_{p,q}\|$ now as given in (2.12). Consider spaces $L_{p,q}$ and $L_{p,r}$, where $0 < p, q, r < \infty$. Assuming that some counterpart of Proposition 3.4(iii) was true, i.e., that the existence of some positive $c > 0$ such that $\mathcal{E}_\mathsf{G}^{L_{p,q}}(t) \leq c\, \mathcal{E}_\mathsf{G}^{L_{p,r}}(t)$ for small $t > 0$ led to $L_{p,q} \hookrightarrow L_{p,r}$, we thus had to verify that there is some $c > 0$ such that for all $0 < p, q, r < \infty$ satisfying $(q/p)^{1/q} \leq c\,(r/p)^{1/r}$ it follows $r \geq q$; see also Figure 10.

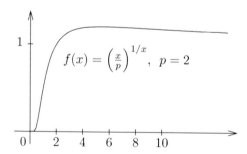

$$f(x) = \left(\frac{x}{p}\right)^{1/x}, \quad p = 2$$

Figure 10

The converse, however, is true: for all $c > 0$ there are p, q, r with $(q/p)^{1/q} \leq c\,(r/p)^{1/r}$ and $r < q$ (given some $c > 0$, choose p with $c > e^{-1/pe}$, $r = pe$, and $q > pe$ sufficiently large).

Let $L_{p,q}\Big([0,\infty)\Big)$ be the Lorentz space with respect to $[\mathcal{R},\mu] = \Big[[0,\infty),|\cdot|\Big]$.

Corollary 3.14 *Let $0 < p, q \le \infty$ (with $q = \infty$ if $p = \infty$). Then*

$$\mathcal{E}_{\mathsf{G}}^{L_{p,q}}(\cdot) \in L_{p,q}\Big([0,\infty)\Big) \qquad \textit{if, and only if,} \qquad q = \infty. \qquad (3.23)$$

P r o o f : This follows immediately from (3.22), Definition (2.12), and Proposition 3.4(i). $\qquad\qquad\square$

We denote by $\mathcal{E}_{\mathsf{G}}^{p,q;a} = \mathcal{E}_{\mathsf{G}}^{L_{p,q}(\log L)_a}(\mathcal{R})$, $0 < p < \infty$, $0 < q \le \infty$, $a \in \mathbb{R}$, for convenience.

Proposition 3.15 *Let $[\mathcal{R},\mu]$ be a σ-finite measure space with $\mathrm{Im}(\mu) = [0,\mu(\mathcal{R})]$ or a finite non-atomic measure space. Then*

$$\mathcal{E}_{\mathsf{G}}^{p,q;a}(t) \sim t^{-\frac{1}{p}} (1 + |\log t|)^{-a}, \qquad 0 < t < \mu(\mathcal{R}), \qquad (3.24)$$

for $0 < p < \infty$, $0 < q \le \infty$, $a \in \mathbb{R}$.

P r o o f : Step 1. First let $q = \infty$; thus $\|f|L_{p,\infty}(\log L)_a\| \le 1$ implies $f^*(t) \le c\, t^{-\frac{1}{p}} (1 + |\log t|)^{-a}$ for any $0 < t < \mu(\mathcal{R})$. Consequently, $\mathcal{E}_{\mathsf{G}}^{p,\infty;a}(t) \le c\, t^{-\frac{1}{p}} (1 + |\log t|)^{-a}$. On the other hand, $L_{p,q}(\log L)_a \hookrightarrow L_{p,\infty}(\log L)_a$, $0 < q \le \infty$, which together with Proposition 3.4(iv) leads to

$$\mathcal{E}_{\mathsf{G}}^{p,q;a}(t) \le c\, t^{-\frac{1}{p}} (1 + |\log t|)^{-a}, \qquad 0 < t < \mu(\mathcal{R}).$$

Step 2. For simplicity we shall first describe the setting in $[\mathcal{R},\mu] = [\Omega, |\cdot|]$, where $\Omega \subseteq \mathbb{R}^n$ is such that it contains $A_s = K_{cs^{1/n}}(0)$. We use the construction (3.14) with $r = p$ and $0 < \varkappa < \frac{1}{p}$; then by Lemma 3.10(ii) $f_{s,\varkappa} \in L_{p,q}(\log L)_a$, $\|f_{s,\varkappa}|L_{p,q}(\log L)_a\| \sim 1$, and

$$\mathcal{E}_{\mathsf{G}}^{p,q;a}(t) \ge \sup_{s>0} f_{s,\varkappa}^*(t) \sim t^{-\varkappa} (1 + |\log t|)^{-a} \sup_{s>t} s^{-\left(\frac{1}{p}-\varkappa\right)}$$

$$\sim t^{-\frac{1}{p}} (1 + |\log t|)^{-a}.$$

This proves (3.24) when $[\mathcal{R},\mu] = [\Omega, |\cdot|]$.

Step 3. In the general case $[\mathcal{R},\mu]$ we can construct the counterpart of $f_{s,\varkappa}$ in (3.14) by some limit procedure arising from simple functions. For $0 < s < \mu(\mathcal{R})$, $m \in \mathbb{N}$, let

$$g_m(x) := \sum_{k=1}^m a_k \chi_{A_k^{m,s}}(x),$$

where we assume that the coefficients satisfy $a_1 > a_2 > \cdots > a_m > 0$, and that the sets $A_k^{m,s}$ are pairwise disjoint subsets of $\mathcal{R}$ with $\bigcup_{k=1}^{m} A_k^{m,s} = A_s$ and $s = \sum_{k=1}^{m} \mu(A_k^{m,s})$. In particular, put

$$a_k \sim [k\mu(A_k^{m,s})]^{-\varkappa} |\log(k\mu(A_k^{m,s}))|^{-a}$$

and $\mu(A_k^{m,s}) \sim \frac{s}{m}$, $k = 1, \ldots, m$. For the monotonicity of $\{a_k\}$ one might have to choose $\varkappa$ properly and s sufficiently small. Then

$$m_k = \sum_{i=1}^{k} \mu(A_i^{m,s}) \sim k\frac{s}{m} \sim k\mu(A_k^{m,s}),$$

and we obtain by Example 2.4 that

$$g_m^*(t) = \sum_{k=1}^{m} a_k \chi_{[m_{k-1},m_k)}(t),$$

that is, for $t \sim k\mu(A_k^{m,s})$,

$$g_m^*(t) \sim a_k \sim [k\mu(A_k^{m,s})]^{-\varkappa} |\log(k\mu(A_k^{m,s}))|^{-a} \sim t^{-\varkappa} |\log t|^{-a}.$$

Now a limit procedure $m \to \infty$ leads to the function $g(x) = \lim_{m\to\infty} g_m(x)$ on A_s, and finally

$$f_{s,\varkappa}(x) := s^{-\left(\frac{1}{p}-\varkappa\right)} g(x)\chi_{A_s}(x), \quad x \in \mathcal{R},$$

is the desired counterpart of (3.14). □

Let $L_{p,q}(\log L)_a\big([0,\varepsilon)\big)$ be the Lorentz-Zygmund space with respect to $[\mathcal{R},\mu] = \big[[0,\varepsilon),|\cdot|\big]$. Then (3.24), Definition (2.13) and Proposition 3.4(i) imply the following counterpart of Corollary 3.14.

Corollary 3.16 *Let $0 < p < \infty$, $0 < q \leq \infty$, $a \in \mathbb{R}$. Then*

$$\mathcal{E}_G^{p,q;a}(\cdot) \in L_{p,q}(\log L)_a\big([0,\varepsilon)\big) \qquad \text{if, and only if,} \qquad q = \infty. \quad (3.25)$$

We deal with the case $p = \infty$ now.

Proposition 3.17 *Let $[\mathcal{R},\mu]$ be a σ-finite measure space with $\mathrm{Im}(\mu) = [0,\mu(\mathcal{R})]$ or a finite non-atomic measure space. Then we obtain for $a \in \mathbb{R}$, $0 < q < \infty$, with $a + 1/q < 0$, and $L_{\infty,q}(\log L)_a = L_{\infty,q}(\log L)_a(\mathcal{R})$,*

$$\mathcal{E}_G^{\infty,q;a}(t) \sim (1 + |\log t|)^{-(a+\frac{1}{q})}, \qquad 0 < t < \varepsilon, \quad (3.26)$$

where $\varepsilon \leq \min(1,\mu(\mathcal{R}))$.

Proof: Step 1. Let $f \in L_{\infty,q}(\log L)_a$ with $\|f|L_{\infty,q}(\log L)_a\| \leq 1$. By (2.13) and the monotonicity of f^* this implies for any number τ, $0 < \tau < \mu(\mathcal{R})$, that

$$f^*(\tau) \left(\int_0^\tau (1 + |\log t|)^{aq} \frac{dt}{t} \right)^{1/q} \leq \left(\int_0^\tau [(1 + |\log t|)^a f^*(t)]^q \frac{dt}{t} \right)^{1/q}$$

$$\leq \left(\int_0^\infty [(1 + |\log t|)^a f^*(t)]^q \frac{dt}{t} \right)^{1/q} \leq 1,$$

i.e., since $a + 1/q < 0$,

$$f^*(\tau) \leq \left(\int_0^\tau (1 + |\log t|)^{aq} \frac{dt}{t} \right)^{-1/q} \leq c (1 + |\log \tau|)^{-(a+\frac{1}{q})}.$$

Step 2. Let $0 < s < 1$, $0 < \varkappa < -(a + \frac{1}{q})$; for simplicity we only describe the setting in $[\mathcal{R}, \mu] = [\mathbb{R}^n, |\cdot|]$. Then by Lemma 3.10(iii),

$$f_{s,\varkappa} \in L_{\infty,q}(\log L)_a, \quad \|f_{s,\varkappa}|L_{\infty,q}(\log L)_a\| \sim 1,$$

with $f_{s,\varkappa}^*(t) \sim (1 + |\log s|)^\varkappa |\log t|^{-(a+\frac{1}{q}+\varkappa)} \chi_{[0,s)}(t)$. Hence,

$$\mathcal{E}_{\mathsf{G}}^{\infty,q;a}(t) \geq \sup_{0<s<1} f_{s,\varkappa}^*(t) \sim |\log t|^{-(a+\frac{1}{q}+\varkappa)} \sup_{t<s<1} (1 + |\log s|)^\varkappa$$

$$= |\log t|^{-(a+\frac{1}{q}+\varkappa)} (1 + |\log t|)^\varkappa \sim (1 + |\log t|)^{-(a+\frac{1}{q})}$$

for $0 < t < 1$. $\qquad\qquad\qquad\qquad\qquad\qquad\qquad\qquad\qquad\qquad \square$

Recall $L_{\infty,\infty}(\log L)_{-a} = L_{\exp,a}$ for $a \geq 0$ and $\mu(\mathcal{R}) < \infty$.

Proposition 3.18 *Let $[\mathcal{R}, \mu]$ be a non-atomic finite measure space, $\mu(\mathcal{R}) = 1$, and $a \geq 0$. Then*

$$\mathcal{E}_{\mathsf{G}}^{L_{\exp,a}}(t) \sim (1 - \log t)^a, \quad 0 < t < 1. \qquad (3.27)$$

Proof: Note that the case $a = 0$ is covered by Proposition 3.12; thus we assume $a > 0$ now. Moreover, by (2.19) we see that $\|f|L_{\exp,a}\| \leq 1$ implies $f^*(t) \leq c (1 - \log t)^a$. Hence $\mathcal{E}_{\mathsf{G}}^{L_{\exp,a}}(t) \leq c (1 - \log t)^a$. Conversely, consider the functions h_s given by Lemma 3.10(iv); thus for $0 < t < 1$,

$$\mathcal{E}_{\mathsf{G}}^{L_{\exp,a}}(t) \geq \sup_{0<s<1} h_s^*(t) \geq c \sup_{s>t} |\log t|^a \sim |\log t|^a,$$

as desired.   □

Let $L_{\exp,a}\left([0,1]\right)$ be the exponential space with respect to $[\mathcal{R},\mu] = \left[[0,1],|\cdot|\right]$.

Corollary 3.19 *Let $a \geq 0$. Then*

$$\mathcal{E}_{\mathsf{G}}^{L_{\exp,a}}(\cdot) \in L_{\exp,a}\left([0,1]\right). \tag{3.28}$$

Proof: This is an immediate consequence of (3.27), (2.19) and Proposition 3.4(i).   □

3.3   Connection with the fundamental function

In rearrangement-invariant function spaces X one has the concept of the "*fundamental function*" φ_X; we now investigate its connection with the growth envelope function $\mathcal{E}_{\mathsf{G}}^X$ as defined above. For convenience we assume in this section that all function spaces are considered over the measure space $[\mathbb{R}^n, \ell_n]$, i.e., $\mathbb{R}^n$ equipped with the (n-dimensional) Lebesgue measure ℓ_n. We closely follow the presentation in [BS88, Ch. 2, §5].

Recall the notion of a Banach function quasi-norm as presented in Definition 2.11. A function quasi-norm ϱ over a measure space $[\mathcal{R},\mu]$ is said to be *rearrangement-invariant*, if $\varrho(f) = \varrho(g)$ for every pair of equi-measurable functions f and g in $\mathcal{M}^+$ that are finite μ-a.e., i.e., if $\mu_f(\lambda) = \mu_g(\lambda)$ for all $\lambda \geq 0$ implies $\varrho(f) = \varrho(g)$. A (quasi-) Banach function space X generated by a rearrangement-invariant (quasi-) norm ϱ is called a *rearrangement-invariant (quasi-) Banach function space* or simply a rearrangement-invariant space. Note that by Definitions 2.11(vi) and 2.12 we have $\chi_A \in X$ for all $A \subset \mathcal{R}$ with $\mu(A) < \infty$.

Definition 3.20 *Let X be a rearrangement-invariant Banach function space over $[\mathbb{R}^n, \ell_n]$. For each $t > 0$, let $A_t \subset \mathbb{R}^n$ be such that $\ell_n(A_t) = t$, and let*

$$\varphi_X(t) = \left\|\chi_{A_t} \,|\, X\right\|. \tag{3.29}$$

The function φ_X so defined is called the fundamental function *of X.*

Note that the particular choice of the set A_t with $\ell_n(A_t) = t$ is immaterial since if B_t is any other subset $B_t \subset \mathbb{R}^n$ with $\ell_n(B_t) = t$, then χ_{A_t} and χ_{B_t} are equi-measurable, and so $\|\chi_{A_t}|X\| = \|\chi_{B_t}|X\|$ because of the rearrangement-invariance of X. Hence φ_X is well-defined. We start with some well-known examples. Let $1 \le p \le \infty$, and $L_p = L_p(\mathbb{R}^n)$; then

$$\varphi_{L_p}(t) = \left\{ \begin{array}{ll} t^{\frac{1}{p}} & , \quad 1 \le p < \infty \\ \chi_{(0,\infty)}(t), & p = \infty \end{array} \right\}, \quad t \ge 0,$$

cf. [BS88, p. 65]. Moreover, when $1 \le q \le p < \infty$ or $p = q = \infty$, then $L_{p,q}$ is rearrangement-invariant and $\varphi_{L_{p,q}}(t) = t^{\frac{1}{p}}$, see [BS88, Ch. 4, Thm. 4.3]. (In view of Remark 2.9 one can further prove that $L_{p,q}$ is a rearrangement-invariant Banach space for $1 < p < \infty$, $1 \le q \le \infty$, or $p = q = \infty$, when f^* in (2.12) is replaced by f^{**}; cf. [BS88, Ch. 4, Thm. 4.6].) Likewise, let $\Omega \subset \mathbb{R}^n$ have finite measure, say, $\ell_n(\Omega) = 1$. Then it is known that $L_1 (\log L)_1 (\Omega)$ and $L_{\exp,1}(\Omega)$ are rearrangement-invariant with fundamental functions

$$\varphi_{L_1(\log L)_1}(t) = t\left(1 + |\log t|\right), \quad 0 < t < 1,$$

and

$$\varphi_{L_{\exp,1}}(t) = (1 + |\log t|)^{-1}, \quad 0 < t < 1,$$

see [BS88, Ch. 4, Thm. 6.4]. So in view of Propositions 3.12, 3.15 and 3.18 the following assertion seems quite reasonable.

Proposition 3.21 *Let X be a rearrangement-invariant Banach function space over $[\mathbb{R}^n, \ell_n]$, and φ_X the corresponding fundamental function. Then*

$$\mathcal{E}_\mathsf{G}^X(t) = \frac{1}{\varphi_X(t)}, \quad t > 0. \tag{3.30}$$

Proof: **Step 1.** Let $t > 0$ and $A_t \subset \mathbb{R}^n$ be such that $\ell_n(A_t) = t$. Then (by our remarks above) $g_t := \chi_{A_t}/\varphi_X(t) \in X$, $\|g_t|X\| = 1$, and

$$g_t^*(s) = \frac{(\chi_{A_t})^*(s)}{\varphi_X(t)} = \frac{\chi_{[0,t)}(s)}{\varphi_X(t)}.$$

Note that by [BS88, Ch. 2, Cor. 5.3] we have $\varphi_X(t) = 0$ if, and only if, $t = 0$. Moreover, φ_X is continuous and increasing; thus

$$\mathcal{E}_\mathsf{G}^X(s) = \sup_{\|f|X\|\le 1} f^*(s) \ge \sup_{t>0} g_t^*(s) = \sup_{t>0} \frac{\chi_{[0,t)}(s)}{\varphi_X(t)} = \sup_{t>s} \frac{1}{\varphi_X(t)} = \frac{1}{\varphi_X(s)}.$$

Step 2. It remains to prove the converse inequality. The rearrangement-invariance of X implies that for any $g \in X'$, X' being the associate space to X, its norm is given by

$$\|g|X'\| = \sup_{\|f|X\|\le 1} \int_0^\infty f^*(s)g^*(s)\mathrm{d}s,$$

see [BS88, Ch. 2, Cor. 4.4]. Let again $t > 0$ and $A_t \subset \mathbb{R}^n$ be such that $\ell_n(A_t) = t$. Hence for $g = \chi_{A_t} \in X'$ (note that X' is also rearrangement-invariant with that norm and thus $\chi_{A_t} \in X'$, too, for any $A_t \subset \mathbb{R}^n$) we obtain

$$\varphi_{X'}(t) = \left\| \chi_{A_t} | X' \right\| = \sup_{\|f|X\|\leq 1} \left(\int_0^t f^*(s)\mathrm{d}s \right) \geq t \sup_{\|f|X\|\leq 1} f^*(t), \qquad (3.31)$$

where we used the monotonicity of $f^*(t)$ again. Thus (3.31) implies $\varphi_{X'}(t) \geq t\, \mathcal{E}_{\mathsf{G}}^X(t)$. On the other hand, [BS88, Ch. 2, Thm. 5.2] provides $\varphi_X(t)\varphi_{X'}(t) = t$, leading to $\mathcal{E}_{\mathsf{G}}^X(t) \leq \frac{1}{\varphi_X(t)}$ for all $t > 0$. $\qquad\square$

Remark 3.22 One can prove a counterpart of Proposition 3.21 when the underlying measure space $[\mathcal{R}, \mu] = [\mathbb{R}^n, \ell_n]$ is replaced by some non-atomic finite measure space $[\mathcal{R}, \mu]$.

Carro, Pick, Soria and Stepanov studied related questions in [CPSS01]; in particular, [CPSS01, Rem. 2.5(ii)] essentially coincides with (3.30), where their function $\varrho_X(t)$ corresponds to $\mathcal{E}_{\mathsf{G}}^X(t)$. Moreover, when X is a rearrangement-invariant Banach function space, there is a counterpart of Proposition 3.4(iii) in [CPSS01, Thm. 2.8(iii)]:

$$X \hookrightarrow L_{q,\infty} \quad \Longleftrightarrow \quad \sup_{t>0} t^{\frac{1}{q}}\, \mathcal{E}_{\mathsf{G}}^X(t) < \infty, \quad 0 < q < \infty.$$

A further property of the fundamental function φ_X is its *quasi-concavity* by which the following is meant: A non-negative function φ defined on $\mathbb{R}_+$ is called *quasi-concave*, if $\varphi(t)$ is increasing on $(0, \infty)$, $\varphi(t) = 0$ if, and only if, $t = 0$, and $\frac{\varphi(t)}{t}$ is decreasing on $(0, \infty)$; see [BS88, Ch. 2, Def. 5.6]. Observe that every non-negative concave function on $\mathbb{R}_+$, that vanishes only at the origin, is quasi-concave; the converse, however, is not true. However, any quasi-concave function φ is equivalent to its least concave majorant $\widetilde{\varphi}$, cf. [BS88, Ch. 2, Prop. 5.10].

Corollary 3.23 *Let X be a rearrangement-invariant Banach function space over $[\mathbb{R}^n, \ell_n]$, put*

$$\psi_{\mathsf{G}}(t) = t\, \mathcal{E}_{\mathsf{G}}^X(t), \quad t > 0. \qquad (3.32)$$

(i)   *The function $\psi_{\mathsf{G}}(t)$ is monotonically increasing in $t > 0$.*

(ii) *If $\lim_{t\downarrow 0} \psi_{\mathsf{G}}(t) = 0$, then $\psi_{\mathsf{G}}(t)$ is equivalent to some concave function for $t > 0$.*

(iii) *The growth envelope function $\mathcal{E}_{\mathsf{G}}^X(t)$ is equivalent to some convex function for $t > 0$.*

P r o o f : Part (i) follows immediately from Proposition 3.21 and a corresponding result for the fundamental function, stating that $\frac{\varphi_X(t)}{t}$ is decreasing, cf. [BS88, Ch. 2, Cor. 5.3].

As for (iii), we know that φ is quasi-concave; thus by the above-mentioned result it is equivalent to some concave function: hence application of Proposition 3.21 yields that $1/\mathcal{E}_G^X(t)$ is equivalent to some concave function for $t > 0$. One verifies that $\mathcal{E}_G^X(t)$ is equivalent to some convex function on $(0, \infty)$ then.

Finally, (ii) is a consequence of (i), Proposition 3.4(i) and the general statements on concave and quasi-concave functions as repeated above. □

Remark 3.24 The question naturally arises whether the rearrangement-invariance of X is really necessary or to what extent this assumption can be weakened (not to mention extensions of $[\mathcal{R}, \mu] = [\mathbb{R}^n, \ell_n]$ at the moment). In all the cases we studied, i.e., spaces of type $L_{p,q}(\log L)_a$ and $A_{p,q}^s$ (postponed to Part II), respectively, we obtain the above-described behaviour of $\mathcal{E}_G^X$ and ψ_G whenever $X \subset L_1^{\text{loc}}$ is satisfied (incorporating in a slight abuse of notation the case of constant functions ψ_G in (i), too; then also $X = L_1$ with $\mathcal{E}_G^X(t) \sim t^{-1}$ and thus $\psi_G(t) \sim 1$ is covered): functions of type

$$\mathcal{E}_G^X(t) \sim t^{-\varkappa} |\log t|^\mu, \quad t > 0 \quad \text{small},$$

with $0 < \varkappa < 1$, $\mu \in \mathbb{R}$, or $\varkappa = 0$, $\mu > 0$, lead to functions $\psi_G(t)$ clearly satisfying Corollary 3.23 (with the above-mentioned extension to $\varkappa = 1$, $\mu \le 0$).

On the other hand, as we did not observe a direct application of (an extended version of) Corollary 3.23 so far, we postpone the study of what happens when X is not rearrangement-invariant.

3.4 Further examples: Sobolev spaces, weighted L_p - spaces

Returning to our starting point, Sobolev's famous embedding result Theorem 2.28, we study (classical) Sobolev spaces W_p^k, $1 \le p < \infty$, $k \in \mathbb{N}$, now. Moreover, as we intend to compare local and global behaviour of $\mathcal{E}_G^X(t)$, i.e., for $0 < t < 1$, or $t \to \infty$, respectively, in Section 10.3, we prepare this a little and briefly deal with some weighted spaces $L_p(w)$. For convenience we retain the setting $[\mathcal{R}, \mu] = [\mathbb{R}^n, \ell_n]$ from the last section and shall always assume $\Omega = \mathbb{R}^n$ unless otherwise stated.

In view of Theorem 2.28(i) we have $W_p^k \hookrightarrow L_\infty$ for $k > \frac{n}{p}$, $1 \le p < \infty$, or $k = n$ and $p = 1$, such that by Proposition 3.4(iii) the corresponding growth envelope function is bounded. Hence we are left to study the following cases now:

$$W_p^k \not\hookrightarrow L_\infty \quad \text{if} \quad \left\{ \begin{array}{l} k < \frac{n}{p} \,, 1 \le p < \infty \\ k = \frac{n}{p} \,, 1 < p < \infty \end{array} \right\}. \tag{3.33}$$

We start with the sub-critical case $k < \frac{n}{p}$.

Proposition 3.25 *Let* $1 \le p < \infty$, $n \ge 2$, $k \in \mathbb{N}_0$, *with* $k < \frac{n}{p}$. *Then*

$$\mathcal{E}_{\mathsf{G}}^{W_p^k}(t) \sim t^{-\frac{1}{p}+\frac{k}{n}}, \quad 0 < t < 1. \tag{3.34}$$

Proof: By $W_p^0 = L_p$ the case $k = 0$ is covered by Proposition 3.12. Moreover, Sobolev's famous embedding result (2.45) immediately yields

$$\mathcal{E}_{\mathsf{G}}^{W_p^k}(t) \le c \, \mathcal{E}_{\mathsf{G}}^{L_{p^*}}(t) \le c' \, t^{-\left(\frac{1}{p}-\frac{k}{n}\right)}, \quad 0 < t < 1,$$

with $\frac{1}{p^*} = \frac{1}{p} - \frac{k}{n}$, applying Propositions 3.4(iv) and 3.12. Hence it remains to prove the converse inequality. Let $R > 0$ and consider functions

$$f_R(x) = R^{k-\frac{n}{p}} \, \psi\left(R^{-1}x\right), \quad x \in \mathbb{R}^n, \tag{3.35}$$

where $\psi(x)$ is some compactly supported C^∞-function in $\mathbb{R}^n$, e.g., as given by Example 2.6,

$$\psi(x) = \left\{ \begin{array}{ll} e^{-\frac{1}{1-|x|^2}} \,, & |x| < 1, \\ 0 & , |x| \ge 1. \end{array} \right. \tag{3.36}$$

At the moment we may restrict ourselves to small R, $0 < R < 1$. Clearly, by the above construction,

$$\mathrm{D}^\alpha f_R(x) = R^{k-\frac{n}{p}} \, R^{-|\alpha|} \left(\mathrm{D}^\alpha \psi\right)\left(R^{-1}x\right), \quad \alpha \in \mathbb{N}_0^n, \tag{3.37}$$

and thus

$$\|\mathrm{D}^\alpha f_R|L_p\| = R^{k-\frac{n}{p}-|\alpha|} \, \|\mathrm{D}^\alpha \psi\left(R^{-1}\cdot\right)|L_p\| = R^{k-|\alpha|} \, \|\mathrm{D}^\alpha \psi|L_p\|. \tag{3.38}$$

Consequently, for $0 < R < 1$,

$$\|f_R|W_p^k\| \le \left(\sum_{|\alpha| \le k} R^{(k-|\alpha|)p} \, \|\mathrm{D}^\alpha \psi|L_p\|^p \right)^{1/p} \le \|\psi|W_p^k\|,$$

and $g_R := \|\psi|W_p^k\|^{-1} f_R \in W_p^k$, $\|g_R|W_p^k\| \le 1$. On the other hand, by (2.9) and Proposition 2.3,

$$g_R^*(t) = \frac{R^{k-\frac{n}{p}}}{\|\psi|W_p^k\|} \, \psi^*\left(R^{-n}t\right). \tag{3.39}$$

Let $0 < t < 1$ and choose $R_0 = R_0(t) = d\,t^{1/n}$ such that $\left(\frac{t}{|\omega_n|}\right)^{1/n} < R_0 < 1$, i.e., $R_0^{-n}t < |\omega_n|$ for appropriate $d > 0$. This finally leads to

$$\mathcal{E}_{\mathsf{G}}^{W_p^k}(t) \;\geq\; \sup_{0<R<1}\; g_R^*(t) \;\geq\; g_{R_0}^*(t) \geq\; c\, R_0^{k-\frac{n}{p}} \;\geq\; c'\, t^{\frac{k}{n}-\frac{1}{p}},$$

completing the proof. $\hfill\square$

Remark 3.26 In Part II when we deal with spaces of Besov and Triebel-Lizorkin type, $B_{p,q}^s$ and $F_{p,q}^s$, respectively, we shall stress arguments similar to those above and "extremal" functions of type (3.35) will recur and emerge as so-called "atoms" in the corresponding spaces. Moreover, slightly adapted functions of type (3.35) will be used to investigate global assertions, see Section 10.3, then for large values of R, $R \to \infty$. Obviously, we have $f \in W_p^k \hookrightarrow L_p$ and hence, by Propositions 3.4(iv) and 3.12 immediately $f^*(t) \leq c\,t^{-\frac{1}{p}}, t > 0$. This results in a worse upper bound than (3.34) for small t (local singularities). In other words, Sobolev's famous embedding theorem (2.45) reads in this context as the conclusion that the increased smoothness assertions imposed on $f \in W_p^k$ (compared with $f \in L_p$ simply) lead to the reduction of admitted local singularities (unboundedness). We shall see in Section 10.3 that this is different from the global behaviour. Moreover, by Sobolev's embedding (2.45) we only know $W_p^k \hookrightarrow L_r, \frac{1}{r} = \frac{1}{p} - \frac{k}{n}$, but the corresponding growth envelope functions $\mathcal{E}_{\mathsf{G}}^{W_p^k}$ and $\mathcal{E}_{\mathsf{G}}^{L_r}$ even coincide – unlike the underlying spaces. We return to this observation in more general context in Remark 8.2.

We deal with the case $k = \frac{n}{p}$, $1 < p < \infty$ now. Clearly, by (2.44) and Proposition 3.4(iii), $\mathcal{E}_{\mathsf{G}}^{W_p^{n/p}}(t)$ cannot be bounded for $t \downarrow 0$. On the other hand, results of Trudinger [Tru67], Yudovich [Yud61], Pohožaev [Poh65], Moser [Mos71], Strichartz [Str72], Maz'ya [Maz72] yield (in our notation)

$$W_p^{n/p}(\Omega) \;\hookrightarrow\; L_{\exp,\frac{1}{p'}}(\Omega), \qquad 1 < p < \infty, \qquad (3.40)$$

for a bounded domain $\Omega \subset \mathbb{R}^n$, say, with $|\Omega| \leq 1$. This leads to the following result.

Proposition 3.27 *Let* $1 < p < \infty$ *be such that* $\frac{n}{p} = k \in \mathbb{N}$. *Then*

$$\mathcal{E}_{\mathsf{G}}^{W_p^{n/p}}(t) \;\sim\; |\log t|^{\frac{1}{p'}}, \qquad 0 < t < \frac{1}{2}. \qquad (3.41)$$

Proof: Let $f \in W_p^{n/p}(\mathbb{R}^n)$. Then we have locally always

$$f^*(t) \leq c \left\| f | W_p^{n/p} \right\| |\log t|^{\frac{1}{p'}}, \quad 0 < t < \frac{1}{2}, \tag{3.42}$$

cf. [BW80, p. 787], so that the upper estimate immediately follows,

$$\mathcal{E}_{\mathsf{G}}^{W_p^{n/p}}(t) \leq c \, |\log t|^{\frac{1}{p'}}, \quad 1 < p < \infty, \quad 0 < t < \frac{1}{2}. \tag{3.43}$$

For the converse, we need a refined version of (3.35), that is,

$$f_m(x) = \sum_{j=0}^{m-1} \psi\left(2^j x\right), \quad m \in \mathbb{N}, \tag{3.44}$$

and ψ given by (3.36). Then by (3.37) and $k = \frac{n}{p}$,

$$(D^\alpha f_m)(x) = \sum_{j=0}^{m-1} 2^{j|\alpha|} (D^\alpha \psi)\left(2^j x\right), \quad m \in \mathbb{N}, \quad \alpha \in \mathbb{N}_0^n,$$

so that for $\alpha \in \mathbb{N}_0^n$, $|\alpha| \leq k$,

$$\begin{aligned}
\|D^\alpha f_m | L_p\| &\leq \left(\sum_{j=0}^{m-1} 2^{j|\alpha|p - jn} \|D^\alpha \psi | L_p\|^p \right)^{1/p} \\
&\leq c \, \|D^\alpha \psi | L_p\| \left(\sum_{j=0}^{m-1} 2^{jp(k - \frac{n}{p})} \right)^{1/p} \\
&= c \, m^{1/p} \|D^\alpha \psi | L_p\|.
\end{aligned}$$

Consequently,

$$\left\| f_m | W_p^{n/p} \right\| \leq c' \, m^{1/p} \left\| \psi | W_p^{n/p} \right\| \leq C \, m^{1/p}, \quad m \in \mathbb{N}. \tag{3.45}$$

On the other hand,

$$f_m^*(t) \sim \begin{cases} m & , \quad t \leq 2^{-mn}, \\ |\log t|, & 2^{-mn} \leq t \leq 1, \end{cases} \tag{3.46}$$

so that finally, for $0 < t < \frac{1}{2}$,

$$\mathcal{E}_{\mathsf{G}}^{W_p^{n/p}}(t) \geq c \, \sup_{m \in \mathbb{N}} m^{-1/p} f_m^*(t) \geq c \, m_0^{-1/p} f_{m_0}^*(t) \geq c' |\log t|^{-\frac{1}{p}+1}$$

where we have chosen $m_0 \in \mathbb{N}$ such that $t \sim 2 \, 2^{-m_0 n}$, i.e., $m_0 \sim |\log t|$. $\square$

Remark 3.28 In Part II, more precisely, in Theorem 8.16(i) we shall prove the counterpart of Proposition 3.27 in a more general setting.

We come to weighted L_p-spaces now. There are essentially two reasonable ways to explain $L_p(w)$ for some positive weight function $w \in L_1^{\text{loc}}$, i.e., $f \in L_p(w) \iff wf \in L_p$, or, alternatively, $f \in L_p(w) \iff w^{1/p}f \in L_p$. We shall stick to the first possibility, mainly because of historical reasons in connection with weighted spaces of Besov or Triebel-Lizorkin type, see Remark 3.29 below. Hence,

$$\|f|L_p(w)\| = \|wf|L_p\| = \left(\int_{\mathbb{R}^n} |f(x)|^p \, w(x)^p \, \mathrm{d}x \right)^{1/p}. \tag{3.47}$$

We start with some "admissible" weights of at most polynomial growth. More precisely, we shall mean by this the collection of all positive C^∞ functions w on $\mathbb{R}^n$ with the following properties:

(i) for all $\gamma \in \mathbb{N}_0^n$ there exists a positive constant c_γ with

$$|\mathrm{D}^\gamma w(x)| \le c_\gamma \, w(x) \quad \text{for all} \quad x \in \mathbb{R}^n,$$

(ii) there exist two constants $c > 0$ and $\alpha \ge 0$ such that for all $x, y \in \mathbb{R}^n$,

$$0 < w(x) \le c \, w(y) \, \langle x - y \rangle^\alpha.$$

As a prototype we may take first

$$w_\alpha(x) = \langle x \rangle^\alpha, \quad \alpha \in \mathbb{R}, \quad x \in \mathbb{R}^n. \tag{3.48}$$

Remark 3.29 The study of such special weights has a little history, in particular in connection with spaces of Besov or Triebel-Lizorkin type as performed in Part II in detail. In [HT94a], [HT94b], [Har95], [Har97] we studied characterisations and compact embeddings of corresponding spaces, see also [ET96, Ch. 4] and [ST87] for a more general approach. Quite recently, there was a renewed interest in this topic leading to the series of papers [KLSS06a],[KLSS06b], [KLSSxx] and [HT05], [Skr05]. Applications are described in [HT94b].

Note that for $\alpha < -\frac{n}{p}$ the spaces $L_p(w_\alpha)$ do not possess a growth envelope function in the sense of Lemma 3.3(i), i.e., $\mathcal{E}_{\mathsf{G}}^{L_p(w_\alpha)}(t)$ is not finite for (small) $t > 0$. This can be seen by taking $f_\beta(x) = \langle x \rangle^\beta \in L_p(w_\alpha)$, $0 < \beta < -\alpha - \frac{n}{p}$, such that $(f_\beta)^*(t)$ does not exist for $t > 0$. Plainly, when $\mathbb{R}^n$ is replaced by some bounded domain $\Omega \subset \mathbb{R}^n$, we have

$$L_p(\Omega, w_\alpha) = L_p(\Omega), \quad \Omega \subset \mathbb{R}^n, \ \Omega \text{ bounded}, \quad \alpha \in \mathbb{R},$$

and all considerations from the unweighted case apply. However, as already mentioned a few times, we are essentially interested in the situation when $\Omega = \mathbb{R}^n$ at this moment.

Proposition 3.30 *Let* $0 < p < \infty$, $\alpha \geq 0$, *then*

$$\mathcal{E}_{\mathsf{G}}^{L_p(w_\alpha)}(t) \sim t^{-\frac{1}{p}}, \quad 0 < t < 1. \tag{3.49}$$

P r o o f : Note that $w_\alpha(x) \geq 1$, $x \in \mathbb{R}^n$, hence

$$L_p(w_\alpha) \hookrightarrow L_p, \quad \alpha \geq 0, \tag{3.50}$$

and Propositions 3.4(iv), 3.12 imply $\mathcal{E}_{\mathsf{G}}^{L_p(w_\alpha)}(t) \leq t^{-\frac{1}{p}}$, $t > 0$. Conversely, using Lemma 3.10(i) with $r = p$, it is sufficient to show that $\|f_s|L_p(w_\alpha)\| \sim 1$ for small s,

$$\|f_s|L_p(w_\alpha)\|^p = s^{-1} \int_{A_s} w_\alpha(x)^p \mathrm{d}x = s^{-1} \int_{A_s} \langle x \rangle^{\alpha p} \mathrm{d}x$$

$$= s^{-1} \int_0^{cs^{1/n}} \left(1 + |x|^2\right)^{\frac{\alpha p}{2}} \mathrm{d}x \leq c'' \, s^{-1} \left(1 + c'\right)^{\frac{\alpha p}{2}} |A_s| \leq C$$

as $0 < s < 1$. In the same way as in the proof of Proposition 3.12 this leads to $\mathcal{E}_{\mathsf{G}}^{L_p(w_\alpha)}(t) \geq t^{-\frac{1}{p}}$, $0 < t < 1$, thus completing the proof. $\square$

Remark 3.31 It is immediately clear that assertion (3.49) remains unchanged when w_α is replaced by an arbitrary admissible weight w that is bounded from below, $w(x) \geq c > 0$, $x \in \mathbb{R}^n$; one can either adapt the above proof appropriately or, even simpler, conclude that due to the admissibility of w, in particular (ii), there are constants $c' > c > 0$ and $\alpha \geq 0$ such that

$$c \leq w(x) \leq c' \langle x \rangle^\alpha, \quad x \in \mathbb{R}^n.$$

Thus

$$L_p(w_\alpha) \hookrightarrow L_p(w) \hookrightarrow L_p, \tag{3.51}$$

and Propositions 3.4(iv), 3.12, 3.30 complete the argument.

Finally, we illuminate another famous class of weights, the Muckenhoupt $\mathcal{A}_p$ weights. Recall that a function $w \in L_1^{\mathrm{loc}}$, $w > 0$ a.e., satisfies the $\mathcal{A}_p$ condition, $1 < p < \infty$, if there is some constant $A > 0$ such that for all balls B in $\mathbb{R}^n$,

$$\frac{1}{|B|} \int_B w(x) \mathrm{d}x \left(\frac{1}{|B|} \int_B w(x)^{-\frac{p'}{p}} \mathrm{d}x \right)^{p/p'} \leq A < \infty.$$

For $p = 1$ the condition reads as $w \in \mathcal{A}_1$ if there is some $A > 0$ such that for all balls B in $\mathbb{R}^n$ and for a.e. $x \in B$,

$$\frac{1}{|B|} \int_B w(x)\mathrm{d}x \leq A\, w(x),$$

and, finally,

$$\mathcal{A}_\infty := \bigcup_{p>1} \mathcal{A}_p.$$

Remark 3.32 The class of such weights has been intensively studied in the past; we refer to [Muc72], [Muc74], and the monographs [GCR85], [Ste93, Ch. V] for a detailed account. Additionally we mention that in [Ryc01] there was introduced a more general concept of weight functions, $\mathcal{A}_p^{\mathrm{loc}}$, containing both the above admissible weights as well as $\mathcal{A}_p$ weights.

Again we content ourselves with an example and consider the counterpart to w_α,

$$w^\alpha(x) = |x|^\alpha, \quad \alpha \in \mathbb{R}. \tag{3.52}$$

It is well-known that $w^\alpha \in \mathcal{A}_p$, $1 < p < \infty$, if, and only if, $-\frac{n}{p} < \alpha < \frac{n}{p'}$, cf. [Ste93, Ch. V, §6.4].

Lemma 3.33 *Let* $1 \leq u \leq \infty$, $1 < p < \infty$, $0 \leq \alpha < \frac{n}{p'}$, *and* p_0 *be given by* $\frac{1}{p_0} = \frac{1}{p} + \frac{\alpha}{n}$. *Let* w *stand for* w^α *or* w_α, *respectively. Then*

$$L_{p,u}(w) \hookrightarrow L_{p_0,u}, \tag{3.53}$$

and, in particular,

$$L_p(w) \hookrightarrow L_{p_0,p}. \tag{3.54}$$

P r o o f : We combine Hölder's inequality and real interpolation arguments for Lorentz spaces,

$$(L_{r_0,q_0}, L_{r_1,q_1})_{\theta,p} = L_{r,p}, \tag{3.55}$$

where $0 < \theta < 1$, $0 < r_0, r_1 < \infty$, $r_0 \neq r_1$, $0 < q_0, q_1, p \leq \infty$, and

$$\frac{1}{r} = \frac{1-\theta}{r_0} + \frac{\theta}{r_1}; \tag{3.56}$$

this is the very classical interpolation result for Lorentz spaces, cf. [BL76, Thm. 5.3.1] and [Tri78a, Thm. 1.18.6/2]. In that way one can show that if $1 < p, q < \infty$ with $0 < \frac{1}{r} = \frac{1}{p} + \frac{1}{q} < 1$, and $1 \leq u \leq \infty$, then

$$L_{p,u} \cdot L_{q,\infty} \subset L_{r,u}, \tag{3.57}$$

in the sense that whenever $f \in L_{p,u}$ and $g \in L_{q,\infty}$, then $fg \in L_{r,p}$; for a short proof see [Har98, Lemma 2.12]. Furthermore, $w^{-1} \in L_{n/\alpha,\infty}$. Thus with $q = \frac{n}{\alpha}$, and thus $r = p_0$ given by $\frac{1}{p_0} = \frac{1}{p} + \frac{\alpha}{n}$; (3.57) implies

$$\left\| f | L_{p_0,u} \right\| = \left\| fww^{-1} | L_{p_0,u} \right\| \leq c \left\| fw | L_{p,u} \right\| \left\| w^{-1} | L_{\frac{n}{\alpha},\infty} \right\| \leq c' \left\| f | L_{p,u}(w) \right\|,$$

i.e., $L_{p,u}(w) \hookrightarrow L_{p_0,u}$. With $u = p$, (3.54) immediately follows. □

Remark 3.34 In Corollary 11.7 we complement the above embedding assertions by sufficient conditions for (3.54).

Proposition 3.35 *Let* $1 < p < \infty$, $0 \leq \alpha < \frac{n}{p'}$; *then*

$$\mathcal{E}_{\mathsf{G}}^{L_p(w^\alpha)}(t) \sim t^{-\frac{\alpha}{n}-\frac{1}{p}}, \quad t > 0. \tag{3.58}$$

P r o o f : The upper estimate is an immediate consequence of (3.54) with $w = w^\alpha$ and Propositions 3.4(iv), 3.12. For the lower one consider extremal functions of type (3.11) with $r = p_0$; thus

$$\left\| f_s | L_p(w^\alpha) \right\|^p = s^{-\frac{p}{p_0}} \int_{A_s} w^\alpha(x)^p \mathrm{d}x \sim s^{-\frac{p}{p_0}} \int_0^{cs^{1/n}} |x|^{\alpha p} \mathrm{d}x \sim s^{-\frac{p}{p_0}} \left(s^{\frac{1}{n}} \right)^{\alpha p+n} \sim c',$$

i.e., (up to possible normalising factors) we have $\left\| f_s | L_p(w^\alpha) \right\| \leq 1$. Hence,

$$\mathcal{E}_{\mathsf{G}}^{L_p(w^\alpha)}(t) \geq \sup_{s>0} f_s^*(t) \geq c \sup_{s>t} s^{-\frac{1}{p_0}} \sim t^{-\frac{\alpha}{n}-\frac{1}{p}}, \quad t > 0,$$

where we additionally used (3.12). □

Remark 3.36 Note that the argument for the lower bound works for all $\alpha > -\frac{n}{p}$. Similarly, by (3.54) we had the counterpart of the upper estimate for w_α, too. But for small $t > 0$ this obviously leads to a weaker estimate than (3.49); however, for large numbers t it is not difficult to predict that – as in the case of w^α in (3.58) – the number p_0 (and thus α) will determine the behaviour of $\mathcal{E}_{\mathsf{G}}^{L_p(w_\alpha)}(t)$, $t \to \infty$. This discussion is postponed to Section 10.3. Comparing Propositions 3.30 and 3.35 it is clear that the influences of the locally regular weight w_α and the $\mathcal{A}_p$-weight w^α, concerning local singularities in the underlying spaces, are different.

As mentioned above, parameters $\alpha \leq -\frac{n}{p}$ are not admitted for w^α or w_α, respectively, whereas the case $-\frac{n}{p} < \alpha < 0$ is reasonable to consider. This will be done elsewhere.

Chapter 4

Growth envelopes $\mathfrak{E}_G$

We shall need a finer characterisation than that provided by the growth envelope functions only. By Proposition 3.15 it is obvious, for instance, that $\mathcal{E}_G^X$ alone cannot distinguish between different spaces like $L_{p,q_1}(\log L)_a$ and $L_{p,q_2}(\log L)_a$. So it is desirable to complement $\mathcal{E}_G^X$ by some expression, naturally belonging to $\mathcal{E}_G^X$, but yielding – as a test – the number q (or a related quantity) in case of $L_{p,q}(\log L)_a$ spaces. Again a more substantial justification for complementing $\mathcal{E}_G^X$ by this additional expression results from more complicated spaces (like $B_{p,q}^s$ and $F_{p,q}^s$) than $L_{p,q}(\log L)_a$; but in these classical cases the outcome can be checked immediately.

The missing link is obtained by the introduction of some "characteristic" index u_G^X, which gives a finer measure of the (local) integrability of functions belonging to X. Moreover, the definition below is also motivated by (sharp) inequalities of type (1.4) with $\varkappa \equiv 1$.

4.1 Definition

We start with some preliminaries. Let ψ be a real continuous monotonically increasing function on the interval $[0,\varepsilon]$ for some small $\varepsilon > 0$. Assume $\psi(0) = 0$ and $\psi(t) > 0$ if $0 < t \le \varepsilon$. Let $\mu_{\log\psi}$ be the associated Borel measure with respect to the distribution function $\log\psi$; if, in addition, ψ is continuously differentiable in $(0,\varepsilon)$ then

$$\mu_{\log\psi}(\mathrm{d}t) = \frac{\psi'(t)}{\psi(t)}\,\mathrm{d}t \tag{4.1}$$

in $(0,\varepsilon)$; cf. [Lan93, p. 285] or [Hal74, §15(9)]. The following result of Triebel [Tri01, Prop. 12.2] is essential for our argument below.

Proposition 4.1

(i)   *Let ψ and $\mu_{\log\psi}$ be as above, and $0 < r_0 \le r_1 < \infty$. Then there are* *numbers $c_2 > c_1 > 0$ such that*

$$\sup_{0 < t < \varepsilon} \psi(t)\, g(t) \;\leq\; c_1 \left(\int_0^\varepsilon [\psi(t)\, g(t)]^{r_1}\, \mu_{\log \psi}(dt) \right)^{1/r_1}$$

$$\leq\; c_2 \left(\int_0^\varepsilon [\psi(t)\, g(t)]^{r_0}\, \mu_{\log \psi}(dt) \right)^{1/r_0} \tag{4.2}$$

for all functions $g(t) \geq 0$, which are monotonically decreasing.

(ii) *Let ψ_1, ψ_2 be two equivalent functions as above and $\mu_{\log \psi_1}$, $\mu_{\log \psi_2}$ the corresponding measures. Assume $0 < r \leq \infty$. Then*

$$\left(\int_0^\varepsilon [\psi_1(t) g(t)]^r\, \mu_{\log \psi_1}(dt) \right)^{1/r} \sim \left(\int_0^\varepsilon [\psi_2(t) g(t)]^r\, \mu_{\log \psi_2}(dt) \right)^{1/r} \tag{4.3}$$

(usual modification if $r = \infty$) for all functions $g(t) \geq 0$, which are monotonically decreasing.

In a slight abuse of notation we shall mean by μ_G the Borel measure associated with a function ψ (as described above and) equivalent to $1/\mathcal{E}_G^X$, where X is some function space satisfying (3.9) and $X \not\hookrightarrow L_\infty$; that is, $\psi(t) \sim 1/\mathcal{E}_G^X(t)$, $0 < t < \varepsilon$. Note that the equivalence class of growth envelope functions $\mathcal{E}_G^X$ of a space X satisfying the assumptions of Lemma 3.8 contains a continuous representative. If $\mathcal{E}_G^X$ is continuously differentiable, then

$$\mu_G(dt) \sim -\frac{\left(\mathcal{E}_G^X \right)'(t)}{\mathcal{E}_G^X(t)}\, dt \tag{4.4}$$

for small $t > 0$. This approach coincides with that presented by Triebel in [Tri01, Sect. 12.1] and [Tri01, Sect. 12.8]. Recall our notation τ_0 in (3.3).

Definition 4.2 *Let $[\mathcal{R}, \mu]$ be a measure space and $X \not\hookrightarrow L_\infty$ some (quasi-) normed function space on $\mathcal{R}$ satisfying the assumptions of Lemma 3.8, and let $\mathcal{E}_G^X$ be the corresponding growth envelope function. Assume $0 < \varepsilon < \tau_0$. The index u_G^X, $0 < u_G^X \leq \infty$, is defined as the infimum of all numbers v, $0 < v \leq \infty$, such that*

$$\left(\int_0^\varepsilon \left[\frac{f^*(t)}{\mathcal{E}_G^X(t)} \right]^v \mu_G(dt) \right)^{1/v} \;\leq\; c\, \|f|X\| \tag{4.5}$$

(with the usual modification if $v = \infty$) holds for some $c > 0$ and all $f \in X$. Then

$$\mathfrak{E}_{\mathsf{G}}(X) = \left(\mathcal{E}_{\mathsf{G}}^X(\cdot), u_{\mathsf{G}}^X \right) \tag{4.6}$$

is called the growth envelope *for the function space X.*

Remark 4.3 It is clear by Proposition 3.4(v) (with $\varkappa \equiv 1$) that (4.5) holds with $v = \infty$ in any case. Thus the question arises whether (depending upon the underlying function space X) there is some smaller v such that (4.5) is still satisfied. Moreover, it is reasonable to ask for the *smallest* parameter v satisfying (4.5) as the corresponding expressions on the left-hand side are monotonically ordered in v by Proposition 4.1(i) with $g = f^*$ and $\psi \sim 1/\mathcal{E}_{\mathsf{G}}^X$.

The number u_{G}^X in Definition 4.2 is defined as the *infimum* of all numbers v satisfying (4.5); however, it is not clear at the moment, whether this infimum is in fact always a minimum. More precisely, one can study the question what assumptions (on the function space X and the underlying measure space) imply that u_{G}^X satisfies (4.5), too. So far we only know that all cases we studied (as presented below) are examples for the latter case (when u_{G}^X happens to be a minimum), but lack a general answer.

Remark 4.4 Note that we explicitly excluded the case $X \hookrightarrow L_\infty$ (in particular, $X = L_\infty$) in Definition 4.2 above. The problem obviously arises from our notation (4.4). One may, however, adopt the (reasonable) opinion that – in case of bounded growth functions $\mathcal{E}_{\mathsf{G}}^X$ (that is, according to Proposition 3.4(iii), when $X \hookrightarrow L_\infty$) – (4.5) is replaced by

$$\sup_{0 < t < \varepsilon} f^*(t) \leq c \, \|f|X\|,$$

for some $c > 0$ and all $f \in X$; thus $u_{\mathsf{G}}^X := \infty$.

The following assertion is not very complicated to prove but quite effective in application later on.

Proposition 4.5 *Let $[\mathcal{R}, \mu]$ be a measure space, and X_i, $i = 1, 2$, (quasi-) normed function spaces on $\mathcal{R}$ with $X_1 \hookrightarrow X_2$. Assume for their growth envelope functions*

$$\mathcal{E}_{\mathsf{G}}^{X_1}(t) \sim \mathcal{E}_{\mathsf{G}}^{X_2}(t), \quad 0 < t < \varepsilon. \tag{4.7}$$

Then we get for the corresponding indices $u_{\mathsf{G}}^{X_i}$, $i = 1, 2$, that

$$u_{\mathsf{G}}^{X_1} \leq u_{\mathsf{G}}^{X_2}. \tag{4.8}$$

Proof: In view of Remark 4.4 we need not assume $X_i \not\hookrightarrow L_\infty$, $i = 1, 2$, explicitly; when $X_2 \hookrightarrow L_\infty$, then $X_1 \hookrightarrow X_2$ in connection with Remark 4.4 implies $u_\mathsf{G}^{X_1} = u_\mathsf{G}^{X_2} = \infty$ which satisfies (4.8), too. Otherwise, when $X_1 \hookrightarrow L_\infty$, then by Proposition 3.4(iii) $\mathcal{E}_\mathsf{G}^{X_1}(t)$ is bounded; hence by (4.7) $\mathcal{E}_\mathsf{G}^{X_2}(t)$ is bounded, too, and another application of Proposition 3.4(iii) yields $X_2 \hookrightarrow L_\infty$ resulting in the preceding argument. Let $f \in X_1 \hookrightarrow X_2$; we conclude by (4.7) together with Proposition 4.1 with $g = f^*$ that

$$\left(\int_0^\varepsilon \left[\frac{f^*(t)}{\mathcal{E}_\mathsf{G}^{X_1}(t)} \right]^{u_\mathsf{G}^{X_2}} \mu_\mathsf{G}^{X_1}(dt) \right)^{1/u_\mathsf{G}^{X_2}} \sim \left(\int_0^\varepsilon \left[\frac{f^*(t)}{\mathcal{E}_\mathsf{G}^{X_2}(t)} \right]^{u_\mathsf{G}^{X_2}} \mu_\mathsf{G}^{X_2}(dt) \right)^{1/u_\mathsf{G}^{X_2}}$$

$$\leq c \, \|f|X_2\| \leq c' \, \|f|X_1\|$$

for all $f \in X_1$. Thus, by definition of $u_\mathsf{G}^{X_1}$ we obtain (4.8). The modifications for the general case (i.e., when the infimum in (4.5) cannot be replaced by a minimum) are obvious. □

Remark 4.6 We give another interpretation of the meaning of (4.5) in terms of sharp embeddings. Assume that $\mathcal{E}_\mathsf{G}^X(t) \sim t^{-\alpha} |\log t|^\mu$ with $\alpha > 0$, $\mu \in \mathbb{R}$, or $\alpha = 0$, $\mu > 0$ (recall the monotonicity of $\mathcal{E}_\mathsf{G}^X$ near 0). Then

$$\mu_\mathsf{G}(dt) \sim \frac{dt}{t} \quad \text{if} \quad \alpha > 0, \qquad \text{and} \qquad \mu_\mathsf{G}(dt) \sim \frac{dt}{t \, |\log t|} \quad \text{if} \quad \alpha = 0,$$

and (4.5) can be reformulated as follows: What is the smallest space of type $L_{\frac{1}{\alpha},v} (\log L)_{-\mu}$ $(\alpha > 0)$ or of type $L_{\infty,v} (\log L)_{-(\mu+\frac{1}{v})}$ $(\alpha = 0)$, respectively, such that X can be embedded into it continuously? Having this idea in mind the results in Section 4.2 are not very astonishing. However, this is only some *interpretation* of (4.5); the definition itself is independent of any scale of Lorentz spaces as target spaces.

4.2 Examples: Lorentz spaces, Sobolev spaces

We give here our main result for those examples already considered in Sections 3.2 and 3.4.

Theorem 4.7 *Let $[\mathcal{R}, \mu]$ be a σ-finite measure space with $\mathrm{Im}(\mu) = [0, \mu(\mathcal{R})]$ or a finite non-atomic measure space.*

(i)   *Let $0 < p, q \leq \infty$ (with $q = \infty$ when $p = \infty$). Then*

$$\mathfrak{E}_\mathsf{G}\left(L_{p,q} \right) = \left(t^{-\frac{1}{p}}, q \right). \tag{4.9}$$

(ii)   *Let* $0 < p < \infty$, $0 < q \le \infty$, *and* $a \in \mathbb{R}$. *Then*

$$\mathfrak{E}_G\left(L_{p,q}(\log L)_a\right) = \left(t^{-\frac{1}{p}} |\log t|^{-a}, \, q\right). \tag{4.10}$$

(iii)   *Let* $0 < q < \infty$, $a \in \mathbb{R}$, *with* $a + \frac{1}{q} < 0$. *Then*

$$\mathfrak{E}_G\left(L_{\infty,q}(\log L)_a\right) = \left(|\log t|^{-(a+\frac{1}{q})}, \, q\right). \tag{4.11}$$

Proof: <u>Step 1.</u>   We begin with (i). Assume first $p < \infty$. We know by Proposition 3.12 that $\mathcal{E}_G^{L_{p,q}}(t) \sim t^{-\frac{1}{p}}$, $t > 0$. Thus it remains to verify the index q according to (4.5); that is, we look for the smallest possible number v such that for some $c > 0$ and all $f \in L_{p,q}$,

$$\left(\int_0^\varepsilon \left[t^{\frac{1}{p}} f^*(t)\right]^v \frac{dt}{t}\right)^{1/v} \le c \, \|f|L_{p,q}\|. \tag{4.12}$$

Note that by (4.4) and $\mathcal{E}_G^{L_{p,q}}(t) \sim t^{-\frac{1}{p}}$ we have $\mu_G(dt) \sim \frac{dt}{t}$ in that case. In view of (2.12) this question can be reformulated as follows: we ask for the smallest possible number v such that $L_{p,q} \hookrightarrow L_{p,v}$ (at least locally). But here it follows immediately that $v \ge q$, i.e., the least admitted number v is q. It remains to deal with the case $p = \infty$, but we may now refer to Remark 4.4.

   <u>Step 2.</u>   We care about (ii). Proposition 3.15 yields

$$\mathcal{E}_G^{p,q;a}(t) \sim t^{-\frac{1}{p}} \left(1 + |\log t|\right)^{-a} \sim t^{-\frac{1}{p}} |\log t|^{-a}, \quad 0 < t < \varepsilon.$$

Together with (4.4) this implies $\mu_G(dt) \sim \frac{dt}{t}$ again. We thus look for possible numbers v such that

$$\left(\int_0^\varepsilon \left[t^{\frac{1}{p}} (1 + |\log t|)^a f^*(t)\right]^v \frac{dt}{t}\right)^{1/v} \le c \, \|f|L_{p,q}(\log L)_a\|$$

for some $c > 0$ and all $f \in L_{p,q}(\log L)_a$. Again we may understand this in the sense that $L_{p,q}(\log L)_a \hookrightarrow L_{p,v}(\log L)_a$ (locally) for some number v. However, this implies $v \ge q$ again, and (4.10) is proved.

   <u>Step 3.</u>   It remains to deal with (iii). We obtain from Proposition 3.17 that $\mathcal{E}_G^{\infty,q;a}(t) \sim (1 + |\log t|)^{-(a+\frac{1}{q})} \sim |\log t|^{-(a+\frac{1}{q})}, \, 0 < t < \varepsilon$. Consequently

$$\mu_G(dt) \sim \frac{1}{1 + |\log t|} \frac{dt}{t}$$

now. Hence we have to determine the smallest number v such that

$$\left(\int_0^\varepsilon \left[(1 + |\log t|)^{a + \frac{1}{q} - \frac{1}{v}} f^*(t) \right]^v \frac{dt}{t} \right)^{1/v} \leq c \, \|f|L_{\infty,q}(\log L)_a\|$$

holds for some $c > 0$ and all $f \in L_{\infty,q}(\log L)_a$. We apply Proposition 2.10 with $r = v$, $b = a + \frac{1}{q} - \frac{1}{v}$, and get $v \geq q$; this proves (4.11).   □

Remark 4.8 As already announced in Remark 4.6, the above results were to be expected in view of the reformulation of (4.5). We thus obtained what we wanted – a method to recover the fine index q in case of Lorentz (-Zygmund) spaces $L_{p,q}(\log L)_a$. Moreover, with the help of Lemma 3.10 one could immediately construct counter-examples to disprove the existence of some $v < p$ satisfying (4.12); similarly for the other cases. For the cases $p, q \leq 1$ we refer to Remark 3.11.

In view of Section 3.3 the question arises naturally whether we can also identify u_{G}^X as some quantity, known for a long time (and in possibly another context) in Banach space theory. By Theorem 4.7 one has to look for expressions only which take the value q when $X = L_{p,q}(\log L)_a$.

Recall $L_{\infty,\infty}(\log L)_{-a} = L_{\exp,a}$ for $a \geq 0$ and $\mu(\mathcal{R}) < \infty$.

Proposition 4.9 *Let $[\mathcal{R}, \mu]$ be a non-atomic finite measure space, $\mu(\mathcal{R}) = 1$, and $a \geq 0$. Then*

$$\mathfrak{E}_{\mathrm{G}}\left(L_{\exp,a} \right) = \left(|\log t|^a, \infty \right). \tag{4.13}$$

Proof: We know by Proposition 3.18 that $\mathcal{E}_{\mathrm{G}}^{L_{\exp,a}}(t) \sim (1 + |\log t|)^a$, $0 < t < 1$. For which numbers v is there some $c > 0$ such that for all $f \in L_{\exp,a}$,

$$\left(\int_0^\varepsilon \left[(1 + |\log t|)^{-(a + \frac{1}{v})} f^*(t) \right]^v \frac{dt}{t} \right)^{1/v} \leq c \sup_{0 < t < 1} (1 + |\log t|)^{-a} f^*(t)$$

is the question that remains. But obviously this leads to $v = \infty$ only, recall Proposition 2.10.   □

Proposition 4.10 *Let $1 \leq p < \infty$, $n \geq 2$, $k \in \mathbb{N}_0$, with $k < \frac{n}{p}$. Then*

$$\mathfrak{E}_{\mathrm{G}}\left(W_p^k \right) = \left(t^{-\frac{1}{p} + \frac{k}{n}}, p \right).$$

Proof: In view of Proposition 3.25 and Definition 4.2 it remains to prove that

$$\left(\int_0^\varepsilon \left[t^{\frac{1}{p} - \frac{k}{n}} \, f^*(t) \right]^v \frac{dt}{t} \right)^{1/v} \le c \, \|f|W_p^k\| \tag{4.14}$$

holds for some $c > 0$ and all $f \in W_p^k$ if, and only if, $v \ge p$. First we show that there cannot be some number $v < p$ that guarantees (4.14) for all $f \in W_p^k$; this is done by contradiction using a refined construction of "extremal" functions based upon (3.35) with $R = 2^{-j}$. Let $\{b_j\}_{j \in \mathbb{N}}$ be a sequence of non-negative numbers, and let $x^0 \in \mathbb{R}^n$, $|x^0| > 4$, such that supp $\psi \left(2^j \cdot -x^0 \right) \cap$ supp $\psi \left(2^r \cdot -x^0 \right) = \emptyset$ for $j \ne r$, $j, r \in \mathbb{N}_0$, and ψ given by (3.36). Then

$$f_b(x) := \sum_{j=1}^\infty 2^{-j(k - \frac{n}{p})} \, b_j \, \psi \left(2^j x - x^0 \right), \quad x \in \mathbb{R}^n, \tag{4.15}$$

belongs to W_p^k for $b \in \ell_p$, and due to the disjointness of the supports,

$$\|f_b|W_p^k\| \le c \left(\sum_{j=1}^\infty b_j^p \right)^{1/p} = c \, \|b|\ell_p\|. \tag{4.16}$$

On the other hand, $f_b^* \left(c \, 2^{-jn} \right) \ge c' \, b_j \, 2^{-j(k - \frac{n}{p})}$, $j \in \mathbb{N}_0$. For convenience we may assume $b_1 = \cdots = b_{J-1} = 0$, where J is suitably chosen such that $2^{-J} \sim \varepsilon$, given by (4.14). Then by monotonicity arguments,

$$\left(\int_0^\varepsilon \left[t^{\frac{1}{p} - \frac{k}{n}} \, f_b^*(t) \right]^v \frac{dt}{t} \right)^{1/v} \sim \left(\sum_{j=J}^\infty \left[2^{-jn(\frac{1}{p} - \frac{k}{n})} \, f_b^* \left(c \, 2^{-jn} \right) \right]^v \right)^{1/v}$$

$$\ge c \left(\sum_{j=J}^\infty \left[2^{-jn(\frac{1}{p} - \frac{k}{n})} b_j \, 2^{-jn(\frac{k}{n} - \frac{1}{p})} \right]^v \right)^{1/v}$$

$$\sim \left(\sum_{j=J}^\infty b_j^v \right)^{1/v}.$$

Together with (4.16) this implies $\|b|\ell_p\| \ge c \, \|b|\ell_v\|$ for arbitrary sequences of non-negative numbers. This obviously requires $v \ge p$. For the converse we may use real interpolation and Sobolev's embedding: by (2.45) we know

$$W_{p_i}^k \hookrightarrow L_{q_i}, \quad \frac{1}{q_i} = \frac{1}{p_i} - \frac{k}{n}, \quad i = 0, 1;$$

on the other hand, for $1 < p < \infty$,

$$\left(W^k_{p_0}, W^k_{p_1}\right)_{\theta,p} = W^k_p \quad \text{and} \quad \left(L_{q_0}, L_{q_1}\right)_{\theta,p} = L_{q,p},$$

where $0 < \theta < 1$, $1 < q_0 < q_1 < \infty$, $1 < p_0 < p_1 < \infty$, $k \in \mathbb{N}$, and

$$\frac{1}{q} = \frac{1-\theta}{q_0} + \frac{\theta}{q_1} \quad \text{and} \quad \frac{1}{p} = \frac{1-\theta}{p_0} + \frac{\theta}{p_1}. \tag{4.17}$$

The W^k_p-interpolation coincides with [Tri78a, 2.4.2/(10)]; the L-part is the very classical interpolation result for Lebesgue spaces, cf. [BL76, Thm. 5.3.1] and [Tri78a, Thm. 1.18.6/2]. Thus

$$W^k_p \hookrightarrow L_{p^*,p}, \quad \frac{1}{p^*} = \frac{1}{p} - \frac{k}{n}, \quad 1 < p < \infty, \tag{4.18}$$

see also [Pee66]. The case $p = 1$ can be incorporated using results by [Alv77b], [Alv77a],

$$\left(\int_0^\infty \left[t^{\frac{1}{p} - \frac{1}{n}} u^*(t)\right]^r \frac{dt}{t}\right)^{\frac{1}{r}} \leq c \left(\int_0^\infty \left[t^{\frac{1}{p}} |\nabla u|^*(t)\right]^r \frac{dt}{t}\right)^{\frac{1}{r}}, \tag{4.19}$$

for $1 \leq r \leq p < n$ and sufficiently smooth $u \in W^1_p$, see also [Tal94, Thm. 4B]. With $p = r = 1$ this corresponds to the case $k = 1 < n$,

$$\int_0^\infty t^{1 - \frac{1}{n}} u^*(t) \frac{dt}{t} \leq c \int_0^\infty |\nabla u|^*(t) dt = \||\nabla u| \, |L_1\| \leq c \|u|W^1_1\|$$

and covers (4.14) with $v = p = 1$ and $k = 1$. We indicate how this result can be iterated to include $k \in \mathbb{N}$, $k \geq 2$. Assume $u \in W^2_p$, $k = 2 < \frac{n}{p}$, u sufficiently smooth; then $|\nabla u| \in W^1_p$, $1 \leq p < \frac{n}{2}$. Moreover, by (an iterated version of) Sobolev's embedding theorem (2.45) we know that $u \in W^2_p \hookrightarrow W^1_{p_*}$ with $p_* := \frac{np}{n-p}$. By our assumptions, $1 < p_* < n$ such that we can apply (4.19) to $u \in W^1_{p_*}$ (with p replaced by p_*) and obtain for $1 \leq r \leq p_* < n$,

$$\left(\int_0^\infty \left[t^{\frac{1}{p_*} - \frac{1}{n}} u^*(t)\right]^r \frac{dt}{t}\right)^{\frac{1}{r}} \leq c \left(\int_0^\infty \left[t^{\frac{1}{p_*}} |\nabla u|^*(t)\right]^r \frac{dt}{t}\right)^{\frac{1}{r}},$$

i.e., with $r = 1$,

$$\int_0^\infty t^{\frac{1}{p} - \frac{2}{n}} u^*(t) \frac{dt}{t} \leq c \int_0^\infty t^{\frac{1}{p} - \frac{1}{n}} |\nabla u|^*(t) \frac{dt}{t}, \tag{4.20}$$

for $1 \leq \frac{np}{n-p} < n$. On the other hand, (4.19) (with $r = 1$) applied to $g = |\nabla u| \in W_p^1$ leads to

$$\int_0^\infty t^{\frac{1}{p}-\frac{1}{n}} \, |\nabla u|^*(t) \frac{dt}{t} \leq c \int_0^\infty t^{\frac{1}{p}} \, |\nabla g|^*(t) \, \frac{dt}{t}, \qquad (4.21)$$

for $1 \leq p < n$. Consequently, (4.20) and (4.21) with $p = 1$ yield

$$\int_0^\infty t^{1-\frac{2}{n}} \, u^*(t) \frac{dt}{t} \leq c \int_0^\infty |\nabla g|^*(t) \, dt \leq c' \, \big\|g|W_1^1\big\| \leq c'' \, \big\|u|W_p^2\big\|;$$

This gives (4.14) with $v = p = 1$ and $k = 2$. Obviously this process can be iterated so as to cover (4.14) with $v = p = 1$ and all $k \in \mathbb{N}$. Alternatively, one may use [KPxx, Thm. 4.2]. A combination of Theorem 4.7(i) and Proposition 4.5 finishes the proof. □

We continue Proposition 3.27 dealing with the case $k = \frac{n}{p}$.

Proposition 4.11 *Let* $1 < p < \infty$ *be such that* $\frac{n}{p} = k \in \mathbb{N}$. *Then*

$$\mathfrak{E}_G\left(W_p^{n/p}\right) = \left(|\log t|^{\frac{1}{p'}}, \, p\right).$$

Proof: In view of Proposition 3.27 and

$$\mu_G(dt) \sim \frac{1}{|\log t|} \frac{dt}{t}, \quad 0 < t < \varepsilon < \frac{1}{2},$$

we have to show that

$$\left(\int_0^\varepsilon \left[|\log t|^{-(\frac{1}{p'}+\frac{1}{v})} f^*(t)\right]^v \frac{dt}{t}\right)^{1/v} \leq c \, \big\|f|W_p^{n/p}\big\|, \quad f \in W_p^{n/p}, \quad (4.22)$$

holds if, and only if, $v \geq p$. Note that by a result of Hansson [Han79, (3.13)] we have in that case (locally)

$$\left(\int_0^\varepsilon \left[\frac{f^*(t)}{|\log t|}\right]^p \frac{dt}{t}\right)^{1/p} \leq c \, \big\|f|W_p^{n/p}\big\|, \qquad (4.23)$$

so that $v \leq p$; see also Maz'ya [Maz72]. For the converse inequality we proceed by contradiction. Assume that (4.22) holds for some $v < p$ and consider slightly modified functions of type (3.44), i.e.,

$$f_{m,b}(x) = \sum_{j=0}^{m-1} b_j \, \psi\left(2^j x\right), \quad m \in \mathbb{N}, \qquad (4.24)$$

where ψ is given by (3.36) and $b_j \geq 0$, $j = 0, \ldots, m - 1$. Then just as in (3.45),

$$\left\| f_{m,b} | W_p^{n/p} \right\| \leq \sum_{|\alpha| \leq k} \| D^\alpha f_{m,b} | L_p \| \leq c \left\| \psi | W_p^{n/p} \right\| \left(\sum_{j=0}^{m-1} b_j^p \right)^{1/p}$$

$$\leq c' \left(\sum_{j=0}^{m-1} b_j^p \right)^{1/p}.$$

If we choose

$$b_j = (j+1)^{-\frac{1}{p}} (1 + \log(j+1))^{-\frac{1}{v}}, \quad j = 0, \ldots, m-1,$$

then

$$\left\| f_{m,b} | W_p^{n/p} \right\| \leq c \left(\sum_{j=1}^{m} \frac{1}{j \, (1 + \log j)^{p/v}} \right)^{1/p} \leq c' \tag{4.25}$$

uniformly in $m \in \mathbb{N}$ because $p > v$. On the other hand, for $r = 1, \ldots, m$,

$$f_{m,b}^* \left(c \, 2^{-rn} \right) \geq c \sum_{j=0}^{r-1} b_j \geq c \, r \, b_{r-1} = c \, r^{\frac{1}{p'}} (1 + \log r)^{-\frac{1}{v}}, \tag{4.26}$$

so that finally (4.22) implies for sufficiently large $m \in \mathbb{N}$,

$$\left\| f_{m,b} | W_p^{n/p} \right\| \geq c_1 \left(\int_0^\varepsilon \left[\frac{f_{m,b}^*(t)}{|\log t|^{\frac{1}{p'} + \frac{1}{v}}} \right]^v \frac{dt}{t} \right)^{\frac{1}{v}} \geq c_2 \left(\sum_{r=1}^{m} \left[\frac{f_{m,b}^*(c \, 2^{-rn})}{r^{\frac{1}{p'} + \frac{1}{v}}} \right]^v \right)^{\frac{1}{v}}$$

$$\geq c_3 \left(\sum_{r=1}^{m} \frac{1}{r \, (1 + \log r)} \right)^{\frac{1}{v}}$$

which does not converge for $m \to \infty$ in contrast to the left-hand side. From this contradiction we conclude $v \geq p$ and the proof is complete. $\qquad\square$

Remark 4.12 This result reappears as a special case from Theorem 8.16(i) below.

Finally we deal with the weighted L_p-spaces as introduced in Section 3.4.

Proposition 4.13 *Let $1 < p < \infty$, $0 < \alpha < \frac{n}{p'}$. Then*

$$\mathfrak{E}_G \left(L_p \left(w_\alpha \right) \right) = \left(t^{-\frac{1}{p}}, \, p \right) \tag{4.27}$$

and

$$\mathfrak{E}_G \left(L_p \left(w^\alpha \right) \right) = \left(t^{-\frac{1}{p} - \frac{\alpha}{n}}, \ p \right). \tag{4.28}$$

P r o o f : In view of Propositions 3.30 and 3.35 we only have to deal with the indices $u_G^{L_p(w_\alpha)}$ and $u_G^{L_p(w^\alpha)}$. Moreover, (3.50), (3.54), Proposition 4.5, and Theorem 4.7(i) imply

$$u_G^{L_p(w_\alpha)} \leq p, \quad u_G^{L_p(w^\alpha)} \leq p.$$

It is thus sufficient to verify the converse inequalities. We start with (4.27). Assume that there is some $v < p$ such that for some $c > 0$ and all $f \in L_p(w_\alpha)$

$$\left(\int_0^\varepsilon \left[t^{\frac{1}{p}} \ f^*(t) \right]^v \frac{dt}{t} \right)^{1/v} \leq c \ \|f|L_p(w_\alpha)\| \tag{4.29}$$

By Lemma 3.10(ii), in particular (3.14), (3.15) with $r = p$, $\varkappa = \frac{1}{p}$, $\gamma \in \mathbb{R}$, we know that for arbitrary $s > 0$, and A_s as in Lemma 3.10(ii),

$$f_{s,\gamma}(x) = |x|^{-\frac{n}{p}} \ (1 + |\log|x||)^{-\gamma} \ \chi_{A_s}(x), \quad x \in \mathbb{R}^n,$$

with

$$f_{s,\gamma}^*(t) \sim t^{-\frac{1}{p}} \ (1 + |\log t|)^{-\gamma} \chi_{[0,s)}(t), \quad t > 0,$$

does not belong to $L_{p,v}$ locally for $\gamma \leq \frac{1}{v}$, as

$$\left(\int_0^\varepsilon \left[t^{\frac{1}{p}} \ f_{s,\gamma}^*(t) \right]^v \frac{dt}{t} \right)^{\frac{1}{v}} \sim \left(\int_0^{\min(\varepsilon, s)} (1 + |\log t|)^{-\gamma v} \frac{dt}{t} \right)^{\frac{1}{v}}$$

diverges for $\gamma v \leq 1$. On the other hand, $f_{s,\gamma} \in L_p(w_\alpha)$ for $\gamma > \frac{1}{p}$ and all $0 < s < 1$,

$$\|f_{s,\gamma}|L_p(w_\alpha)\|^p = \int_{A_s} |x|^{-n} \ (1 + |\log|x||)^{-\gamma p} \ \langle x \rangle^{\alpha p} dx$$

$$\leq c \int_0^1 (1 + \log r)^{-\gamma p} \ \frac{dr}{r} \leq c'.$$

Thus choosing $\frac{1}{p} < \gamma \leq \frac{1}{v}$ and $0 < s < 1$, $f_{s,\gamma}$ disprove (4.29) for $v < p$. Similarly we proceed in case of w^α where (4.29) is replaced now by

$$\left(\int_0^\varepsilon \left[t^{\frac{1}{p} + \frac{\alpha}{n}} \ f^*(t) \right]^v \frac{dt}{t} \right)^{1/v} \leq c \ \|f|L_p(w^\alpha)\|. \tag{4.30}$$

Accordingly we take as counter-examples

$$g_{s,\gamma}(x) = |x|^{-\frac{n}{p}-\alpha} \left(1 + |\log|x||\right)^{-\gamma} \chi_{A_s}(x), \quad x \in \mathbb{R}^n,$$

with

$$g^*_{s,\gamma}(t) \sim t^{-\frac{1}{p}-\frac{\alpha}{n}} \left(1 + |\log t|\right)^{-\gamma} \chi_{[0,s)}(t), \quad t > 0,$$

due to Lemma 3.10(ii), that is (3.14), (3.15) with $r = p$, $\varkappa = \frac{1}{p} + \frac{\alpha}{n}$, $\gamma \in \mathbb{R}$.
In the same way as above we find that

$$\left(\int_0^\varepsilon \left[t^{\frac{1}{p}+\frac{\alpha}{n}} \, g^*_{s,\gamma}(t)\right]^v \frac{dt}{t}\right)^{\frac{1}{v}} \sim \left(\int_0^{\min(\varepsilon,s)} \left(1 + |\log t|\right)^{-\gamma v} \frac{dt}{t}\right)^{\frac{1}{v}}$$

diverges for $\gamma \leq \frac{1}{v}$, whereas for small s, $0 < s < 1$, and $\gamma > \frac{1}{p}$,

$$\|g_{s,\gamma}|L_p(w^\alpha)\|^p = \int_{A_s} |x|^{-n-\alpha p} \left(1 + |\log|x||\right)^{-\gamma p} |x|^{\alpha p} dx$$

$$\leq c \int_0^1 \left(1 + \log r\right)^{-\gamma p} \frac{dr}{r} \leq c',$$

so that the $g_{s,\gamma}$ violate condition (4.30) for appropriately chosen numbers s
and γ when $v < p$. □

Remark 4.14 In view of Remark 3.31 and our arguments above for $w_\alpha(x)$
which can be immediately extended to arbitrary *admissible* weights that are
bounded from below, $w(x) \geq c > 0$, $x \in \mathbb{R}^n$, we are led to

$$\mathfrak{E}_G\left(L_p(w)\right) = \left(t^{-\frac{1}{p}}, p\right)$$

for all such weights w and $0 < p < \infty$. Note that (3.51), Proposition 4.5,
Theorem 4.7(i), Proposition 4.13 directly imply this result. In other words,
the local regularity of such weights implies that the (local) growth envelopes
(i.e., the characterisation of local singularities) remain unchanged compared
with an unweighted L_p-space ($w \equiv 1$). This is essentially different from the
corresponding situation for $\mathcal{A}_p$-weights where we only treated $w^\alpha(x)$ as an
example so far; however, the situation also changes when regarding global
assertions instead of local ones, see Section 10.3 below.

Chapter 5

The continuity envelope function $\mathcal{E}_C$

As in the case of growth envelopes we first introduce the continuity envelope function $\mathcal{E}_C^X$, derive some elementary properties, and discuss examples afterwards. In addition, we prove a certain lift property that will be essential in later applications. In the sequel we deal with $[\mathcal{R}, \mu] = [\mathbb{R}^n, \ell_n]$ only and regard (quasi-) Banach spaces X of functions on $\mathbb{R}^n$.

5.1 Definition and basic properties

This approach is based on the concept of the modulus of continuity and related to questions of Lipschitz continuity; we refer to Section 2.3 for notation and basics.

Definition 5.1 *Let* $X \hookrightarrow C$ *be a function space on* $\mathbb{R}^n$. *The* continuity envelope function $\mathcal{E}_C^X : (0, \infty) \to [0, \infty)$ *is defined by*

$$\mathcal{E}_C^X(t) := \sup_{\|f|X\| \le 1} \frac{\omega(f, t)}{t}, \quad t > 0. \tag{5.1}$$

Remark 5.2 An adapted version of Remark 3.2 holds here, too, concerning the equivalence classes of continuity envelope functions as well as the question of local (instead of global) behaviour of functions, implying our restriction to function spaces on $\mathbb{R}^n$ rather than function spaces on domains. We do not want to repeat the arguments in detail.

First we collect a few elementary properties of $\mathcal{E}_C^X(t)$. Note that $\mathcal{E}_C^X(t)$ cannot be too small as $t \downarrow 0$, for $\mathcal{E}_C^X(t) \searrow 0$ as $t \downarrow 0$ implies that any element of X is constant. Furthermore, one introduces a number τ_0^C – parallel to (3.3) – by

$$\tau_0^C = \tau_0^C(X) := \sup \left\{ t > 0 : \mathcal{E}_C^X(t) > 0 \right\}. \tag{5.2}$$

75

However, as $\mathcal{E}_C^X(t) = 0$ for some $t > 0$ means $\omega(f,t) = 0$ for all $f \in X$ (i.e., elements of X are constant) we are mainly interested in spaces X with $\tau_0^C(X) = \infty$; investigating the local behaviour (small $t > 0$) at the moment, it was even sufficient to assume, say, $\sup\{0 < t < 1 \ : \ \mathcal{E}_C^X(t) > 0\} = 1$.

Proposition 5.3 *Let $X \hookrightarrow C$ be a function space on $\mathbb{R}^n$.*

(i) $\mathcal{E}_C^X$ *is right-continuous and "essentially monotonically decreasing", that is, $\mathcal{E}_C^X$ is equivalent to some monotonically decreasing function.*

(ii) *We have $X \hookrightarrow \mathrm{Lip}^1$ if, and only if, $\mathcal{E}_C^X(\cdot)$ is bounded, i.e., $\sup\limits_{t>0} \mathcal{E}_C^X(t) = \limsup\limits_{t\downarrow 0} \mathcal{E}_C^X(t)$ is finite. In that case it holds*

$$\mathcal{E}_C^X(0) := \limsup_{t\downarrow 0} \mathcal{E}_C^X(t) = \left\| \mathrm{id} : X \to \mathrm{Lip}^1 \right\|.$$

(iii) *Let $X_i \hookrightarrow C$, $i = 1,2$, be function spaces on $\mathbb{R}^n$. Then $X_1 \hookrightarrow X_2$ implies that there is a positive constant c such that for all $t > 0$,*

$$\mathcal{E}_C^{X_1}(t) \leq c\, \mathcal{E}_C^{X_2}(t).$$

One may choose $c = \|\mathrm{id} : X_1 \to X_2\|$ in that case.

(iv) *Let $X \hookrightarrow C$ be non-trivial, i.e., $\tau_0^C(X) = \infty$. Let $\varkappa : (0,\infty) \to [0,\infty)$ be some non-negative function. Then $\varkappa(\cdot)$ is bounded if, and only if, there is some $c > 0$ such that for all $f \in X$, $\|f|X\| \leq 1$,*

$$\sup_{t>0} \frac{\varkappa(t)}{\mathcal{E}_C^X(t)} \frac{\omega(f,t)}{t} \leq c. \tag{5.3}$$

Proof: Step 1. The continuity of $\mathcal{E}_C^X$ follows immediately from the corresponding feature of $\omega(f,t)$, see [DL93, Ch. 2, §6]. Similarly we gain from a result of DeVore and Lorentz which provides that

$$\frac{1}{2}\,\overline{\omega}(f,t) \leq \omega(f,t) \leq \overline{\omega}(f,t), \quad t > 0, \tag{5.4}$$

for any $f \in C$; cf. [DL93, Ch. 2, Lemma 6.1]. Here $\overline{\omega}(f,t)$ is the least concave majorant of $\omega(f,t)$ (being itself a modulus of continuity). Consequently, as $\overline{\omega}(f,t)$ is concave, $\frac{\overline{\omega}(f,t)}{t}$ is monotonically decreasing in $t > 0$. But this implies (i).

Step 2. We prove (ii). Assume first $X \hookrightarrow \mathrm{Lip}^1$. Then there is some $c > 0$ such that for all $f \in X$, $\|f|\mathrm{Lip}^1\| \leq c\,\|f|X\|$. Thus for all $f \in X$, $\|f|X\| \leq 1$, we obtain $\|\frac{1}{c}f|\mathrm{Lip}^1\| \le 1$. Hence, with $g := \frac{1}{c}f$, $f \in X$, $\|f|X\| \le 1$ implies $g \in \mathrm{Lip}^1$, $\|g|\mathrm{Lip}^1\| \le 1$ and $\omega(f,t) = c\,\omega(g,t)$. Consequently,

$$\mathcal{E}_C^X(t) = \sup_{\|f|X\|\le1} \frac{\omega(f,t)}{t} \le c \sup_{\|g|\mathrm{Lip}^1\|\le1} \frac{\omega(g,t)}{t} \le c \sup_{\|g|\mathrm{Lip}^1\|\le1} \|g|\mathrm{Lip}^1\| \le c.$$

Thus $\mathcal{E}_C^X(t)$ is bounded, and by (i), $\sup_{t>0} \mathcal{E}_C^X(t) \sim \limsup_{t\downarrow0} \mathcal{E}_C^X(t)$ is finite. Conversely, assume that there exists $C > 0$ such that $\sup_{t>0} \mathcal{E}_C^X(t) \le C$. Then

$$\sup_{t>0} \mathcal{E}_C^X(t) = \sup_{t>0} \sup_{\|f|X\|\le1} \frac{\omega(f,t)}{t} = \sup_{\|f|X\|\le1} \sup_{t>0} \frac{\omega(f,t)}{t} \le C,$$

which implies that for any $f \in X$, $\|f|X\| \le 1$, we obtain $\sup_{t>0} \frac{\omega(f,t)}{t} \le C$. Together with our assumption $X \hookrightarrow C$ we can conclude that $f \in \mathrm{Lip}^1$ for any $f \in X$, $\|f|X\| \le 1$; but by some scaling argument this implies $X \hookrightarrow \mathrm{Lip}^1$, as desired. The proof of (iii) is similar.

Step 3. The restriction in (iv) that X contains essentially more than constants only implies that $\mathcal{E}_C^X(t) > 0$ for any $t > 0$. So the left-hand side of (5.3) is well-defined. Now one may proceed analogously to the proof of Proposition 3.4(v). $\square$

Corollary 5.4 *Let* $X \hookrightarrow C$ *be non-trivial,* $\varphi : (0,\infty) \to (0,\infty)$ *a positive, monotonically decreasing function with the following property: For any nonnegative function* $\varkappa : (0,\infty) \to [0,\infty)$ *it is the case, that* $\varkappa(\cdot)$ *is bounded if, and only if, there is some* $c > 0$ *such that for all* $f \in X$, $\|f|X\| \le 1$,

$$\sup_{t>0} \frac{\varkappa(t)}{\varphi(t)} \frac{\omega(f,t)}{t} \le c. \tag{5.5}$$

Then $\varphi \sim \mathcal{E}_C^X$, *i.e., there are numbers* $c_2 > c_1 > 0$ *such that for all* $t > 0$,

$$c_1\,\varphi(t) \le \mathcal{E}_C^X(t) \le c_2\,\varphi(t). \tag{5.6}$$

Proof: The proof is similar to that of Corollary 3.6 and will not be repeated here. $\square$

Remark 5.5 In analogy to Remark 3.7 we mention the *continuity function* ψ considered by Triebel with the additional assumptions that $X = B_{p,q}^s$ or $X = F_{p,q}^s$, it behaves like $\psi \sim 1/\mathcal{E}_C^X$.

Remark 5.6 In view of Section 3.1, in particular Lemma 3.3, one may ask whether any space X of the above type possesses a continuity envelope function $\mathcal{E}_C^X$, that is, whether in any admissible situation $\mathcal{E}_C^X(t)$ is finite for any $t > 0$. In contrast to $\mathcal{E}_G^X$, see Lemma 3.3, our assumption $X \hookrightarrow C$ already implies

$$\mathcal{E}_C^X(t) = \sup_{\|f|X\| \leq 1} \frac{\omega(f,t)}{t} \leq \sup_{\|f|X\| \leq 1} \frac{2\,\|f|C\|}{t} \leq 2\,\|\mathrm{id} : X \to C\|\,\frac{1}{t}, \quad t > 0,$$

i.e., there is some $c > 0$ such that for all $t > 0$, $\mathcal{E}_C^X(t) \leq \frac{c}{t}$. In that sense any space $X \hookrightarrow C$ has a continuity envelope function $\mathcal{E}_C^X$.

Corollary 5.7 *Let $X \hookrightarrow C$ be a function space over $[\mathbb{R}^n, \ell_n]$; put*

$$\psi_C(t) = t\,\mathcal{E}_C^X(t), \quad t > 0. \tag{5.7}$$

(i) *The function $\psi_C(t)$ is monotonically increasing in $t > 0$, $\lim\limits_{t\downarrow 0} \psi_C(t) = 0$.*

(ii) *The function $\psi_C(t)$ is equivalent to some concave function for $t > 0$.*

P r o o f :   Note at first, that when X is trivial in the sense, that it contains constants only, we always have $\mathcal{E}_C^X \equiv 0$ and there is nothing to prove. We come to the more interesting case now, when X is not trivial in the above sense. Part (i) follows immediately from the fact that $\omega(f,t)$ is monotonically increasing in $t > 0$ for any $f \in X$, and $\lim\limits_{t\downarrow 0} \omega(f,t) = \omega(f,0) = 0$ for any $f \in X$. Moreover, $\psi_C(t) = 0$ if, and only if, $t = 0$. Together with (i) and Proposition 5.3(i) we thus obtain that $\psi_C(t)$ is equivalent to some quasi-concave function; recall our remarks in front of Corollary 3.23. Another application of [BS88, Ch. 2, Prop. 5.10] finishes the proof.   □

Remark 5.8 The coincidences as well as differences between Corollaries 3.23 and 5.7 are obvious. Note that in all cases we studied we have the counterpart of Corollary 3.23(iii), too, i.e., $\mathcal{E}_C^X$ is (equivalent to) a convex function.
More important from our point of view, however, is the observation that obviously the (different) envelope functions $\mathcal{E}_G^X$ and $\mathcal{E}_C^X$ show similar behaviour; we merely take it as some kind of (delayed) justification that the definition of the two envelope functions – arising in completely different problems when measuring smoothness or unboundedness, respectively – led to parallel concepts, though each one of them separately was initially motivated by suitable classical settings only.

As already mentioned for growth envelopes, we shall work on function spaces X such that there is a continuous representative in the equivalence class of $\mathcal{E}_C^X$, see Lemma 3.8. In particular, for our purpose it was sufficient to obtain the counterpart of (3.8). We give some sufficient condition.

Lemma 5.9 *Assume that X additionally satisfies*

$$\left\| f\left(2^{-1}\cdot\right) | X \right\| \;\leq\; c \; \|f|X\| \tag{5.8}$$

for some $c > 0$ and all $f \in X$. Then

$$\mathcal{E}_C^X\left(2^{-j}\right) \;\sim\; \mathcal{E}_C^X\left(2^{-j+1}\right) \tag{5.9}$$

for some $j_0 \in \mathbb{N}$ and all $j \geq j_0$.

P r o o f : This generalises [Tri01, (12.78)]; see (5.4) and the similar argument following Proposition 3.4. $\square$

5.2 Some lift property

Before we present first examples we shall prepare later considerations by a "lifting" assertion that turns out to be an essential key in the sequel. It provides a relation between the modulus of continuity of a (sufficiently smooth) function and the non-increasing rearrangement of its gradient. The idea is to gain from results obtained in spaces of (sub-) critical type (and hence in terms of growth envelopes) when dealing with (super-) critical spaces (and continuity envelopes). Roughly speaking, we want to "lift" our (sub-) critical results by smoothness 1 to (super-) critical ones. This is at least partly possible. We return to this point later in Section 11.2 and discuss it in more detail. Recall $(\nabla f)(x) = \left(\frac{\partial f}{\partial x_1}(x), \ldots, \frac{\partial f}{\partial x_n}(x)\right)$, $x \in \mathbb{R}^n$, with

$$|\nabla f(x)| = \left(\sum_{l=1}^{n}\left|\frac{\partial f}{\partial x_l}(x)\right|^2\right)^{1/2} \sim \sum_{l=1}^{n}\left|\frac{\partial f}{\partial x_l}(x)\right|. \tag{5.10}$$

Proposition 5.10

(i) *There exists $c > 0$ such that for all $t > 0$ and all $f \in C^1$,*

$$\omega(f,t) \;\leq\; c \int_0^{t^n} s^{\frac{1}{n}-1}\, |\nabla f|^*(s)\mathrm{d}s \;\sim\; \int_0^{t} |\nabla f|^*(\sigma^n)\,\mathrm{d}\sigma. \tag{5.11}$$

(ii) *Let $0 < r \leq \infty$, $u > \frac{1}{r}$, and $0 < \varepsilon < 1$. Then there is a number $c > 0$ such that*

$$\int_0^{\varepsilon}\left[\frac{\omega(f,t)}{t\,|\log t|^u}\right]^r \frac{\mathrm{d}t}{t} \;\leq\; c \int_0^{\varepsilon}\left[\frac{|\nabla f|^*(t)}{|\log t|^u}\right]^r \frac{\mathrm{d}t}{t} \tag{5.12}$$

(with the obvious modification when $r = \infty$) for all $f \in C^1$.

(iii) *Let* $0 < r \leq \infty$, $0 < \varkappa < 1$, *and* $0 < \varepsilon < 1$. *Then there is a number* $c > 0$ *such that*

$$\int_0^\varepsilon \left[\frac{\omega(f,t)}{t^\varkappa} \right]^r \frac{dt}{t} \leq c \int_0^\varepsilon \left[t^{\frac{1}{n}(1-\varkappa)} \, |\nabla f|^* (t) \right]^r \frac{dt}{t} \tag{5.13}$$

(with the obvious modification when $r = \infty$*) for all* $f \in C^1$.

P r o o f : Step 1.   We prove (i). We thank Professor V. Kolyada for the idea for this estimate (5.11). There is a similar observation in [DS84].

Let $x, h \in \mathbb{R}^n$; denote by Q a (closed) cube containing x and $x + h$ with side-length $\ell = |h|$ and sides parallel to the coordinate planes. We estimate

$$|f(x) - f(x+h)| \leq |f(x) - f_Q| + |f_Q - f(x+h)|, \tag{5.14}$$

where

$$f_Q = \frac{1}{|Q|} \int_Q f(y) dy.$$

It is sufficient now to deal with the first term on the right-hand side of (5.14), $|f(x) - f_Q|$ as the second one can be handled similarly; thus

$$|f(x) - f_Q| \leq \frac{1}{|Q|} \int_Q |f(x) - f(y)| dy = \frac{1}{|Q_0|} \int_{Q_0} |f(x) - f(x+\xi)| \, d\xi,$$

where Q_0 is simply the translation of Q such that one corner is allocated in the origin, $|Q_0| = |Q|$. Furthermore,

$$|f(x) - f(x+\xi)| = \left| \int_0^1 \nabla f(x+\tau\xi) \cdot \xi d\tau \right| \leq |\xi| \int_0^1 |(\nabla f)(x+\tau\xi)| \, d\tau$$

$$\leq \ell \int_0^1 |(\nabla f)(x+\tau\xi)| \, d\tau.$$

Taking into account that $|Q_0| = |Q| = \ell^n$ we thus arrive at

$$|f(x) - f_Q| \leq c \, \ell^{-(n-1)} \int_0^1 d\tau \int_{Q_0} |(\nabla f)(x+\tau\xi)| \, d\xi.$$

A change of variable $z = \tau\xi$ in the latter integral yields

$$|f(x) - f_Q| \leq c \, \ell^{-(n-1)} \, \tau^{-n} \int_0^1 d\tau \int_{Q_\tau} |(\nabla f)(x+z)| \, dz,$$

where Q_τ denotes the dilated cube with side length $\tau\ell$. In view of $|Q_\tau| = \tau^n \ell^n$ and

$$\frac{1}{|Q_\tau|} \int\limits_{Q_\tau} |(\nabla f)\,(x+z)|\,dz \ \sim\ |\nabla f|^{**}\,(\tau^n \ell^n)$$

we obtain

$$|f(x) - f_Q| \ \leq\ c\,\ell \int\limits_0^1 |\nabla f|^{**}\,(\tau^n \ell^n)\,d\tau \ =\ c' \int\limits_0^{\ell^n} s^{\frac{1}{n}-1} |\nabla f|^{**}\,(s)ds.$$

By our introductory remarks and standard arguments this implies

$$\omega(f,t) \ =\ \sup_{|h|\leq t}\ \sup_{x\in\mathbb{R}^n}\ |f(x) - f(x+h)| \ \leq\ C \int\limits_0^{t^n} s^{\frac{1}{n}-1}|\nabla f|^{**}\,(s)ds. \quad (5.15)$$

In view of (5.11) it remains to replace $|\nabla f|^{**}$ by $|\nabla f|^*$ in (5.15). Assume first $n > 1$; then we can apply Hardy's inequality (see, for instance, [BS88, Ch. 3, Lemma 3.9]) to the right-hand side of (5.15) and obtain (5.11). In case of $n = 1$ we prove it directly. Let $f \in C^1(\mathbb{R})$; for simplicity we assume $f(0) = 0$. Let $t > 0$, and $0 < x \leq t$; then

$$|f(x)| \ \leq\ \int\limits_0^x |f'(\tau)|d\tau \ \leq\ \int\limits_0^x |f'|^*(s)ds \ =\ x|f'|^{**}(x) \ \leq\ t|f'|^{**}(t)$$

for any $x \leq t$; for the second estimate see, for instance, [BS88, Ch. 2, Thm. 2.2]. Thus

$$\frac{\omega(f,t)}{t} \ \leq\ c\,|f'|^{**}\,(t), \quad 0 < t < \varepsilon, \quad f \in C^1(\mathbb{R}), \quad (5.16)$$

the modification in the general case $(f(0) \neq 0)$ being obvious. But this is (5.11) in the case $n = 1$.

Step 2.   We have to verify (5.12) and start with the case $r = \infty$, that is we show

$$\sup_{0<t<\varepsilon} \frac{\omega(f,t)}{t\,|\log t|^u} \ \leq\ c \sup_{0<t<\varepsilon} \frac{|\nabla f|^*\,(t)}{|\log t|^u} \quad (5.17)$$

for $u > 0$. By (5.11) we have

$$\frac{\omega(f,t)}{t\,|\log t|^u} \ \leq\ \frac{c}{t\,|\log t|^u} \int\limits_0^{t^n} s^{\frac{1}{n}-1}|\nabla f|^*\,(s)ds$$

$$\leq\ c \sup_{0<s<\varepsilon} \frac{|\nabla f|^*\,(s)}{|\log s|^u} \, \frac{\int\limits_0^{t^n} s^{\frac{1}{n}-1}|\log s|^u ds}{t\,|\log t|^u} \ \leq\ c' \sup_{0<s<\varepsilon} \frac{|\nabla f|^*\,(s)}{|\log s|^u}$$

for small $t > 0$. Thus (5.17) is shown.

Step 3.   Assume now $r < \infty$, $u > \frac{1}{r}$; we verify (5.12). Application of (5.11) and a change of variable $s = \sigma^n$ leads to

$$\int_0^\varepsilon \left[\frac{\omega(f,t)}{t \, |\log t|^u} \right]^r \frac{dt}{t} \leq c \int_0^\varepsilon \left[\frac{1}{t \, |\log t|^u} \int_0^t |\nabla f|^* (\sigma^n) \, d\sigma \right]^r \frac{dt}{t}. \qquad (5.18)$$

Assume that $\varepsilon \sim 2^{-J}$, put $t = 2^{-j}$, $j \geq J$, and $\sigma = 2^{-\ell}$, $\ell \geq j$. Then in view of (5.18) and (5.12) it is sufficient to prove

$$\sum_{j=J}^\infty \left[2^j j^{-u} \sum_{\ell=j}^\infty 2^{-\ell} |\nabla f|^* \left(2^{-\ell n} \right) \right]^r \leq C \sum_{k=J}^\infty \left[k^{-u} \, |\nabla f|^* \left(2^{-kn} \right) \right]^r. \qquad (5.19)$$

The left-hand side can be manipulated into

$$\sum_{j=J}^\infty \left[\sum_{\ell=j}^\infty 2^{-(\ell-j)} j^{-u} |\nabla f|^* \left(2^{-\ell n} \right) \right]^r = \sum_{j=J}^\infty \left[\sum_{i=0}^\infty 2^{-i} \, j^{-u} \, |\nabla f|^* \left(2^{-(i+j)n} \right) \right]^r.$$

With $\varrho < r$, we can estimate further

$$\sum_{j=J}^\infty \left[\sum_{\ell=j}^\infty 2^{-(\ell-j)} j^{-u} |\nabla f|^* \left(2^{-\ell n} \right) \right]^r \leq c \sum_{j=J}^\infty \sum_{i=0}^\infty 2^{-i\varrho} j^{-ur} \left[|\nabla f|^* \left(2^{-(i+j)n} \right) \right]^r;$$

in case of $r \leq 1$ this is simply monotonicity, otherwise we need $\varrho < r$ and apply Hölder's inequality. Changing indices as well as the order of summation provides

$$\sum_{j=J}^\infty \sum_{i=0}^\infty 2^{-i\varrho} \, j^{-ur} \left[|\nabla f|^* \left(2^{-(i+j)n} \right) \right]^r$$

$$= \sum_{k=J}^\infty \left[|\nabla f|^* \left(2^{-kn} \right) \right]^r k^{-ur} \sum_{\nu=0}^{k-J} 2^{-\nu\varrho} \left(\frac{k}{k-\nu} \right)^{ur}.$$

Note that $\frac{k}{k-\nu} = 1 + \frac{\nu}{k-\nu} \leq 1 + \nu$ because $\nu \leq k - J \leq k - 1$. Thus the inner sum converges independently of k,

$$\sum_{\nu=0}^{k-J} 2^{-\nu\varrho} \left(\frac{k}{k-\nu} \right)^{ur} \leq \sum_{\nu=0}^{k-J} 2^{-\nu\varrho} \, (1+\nu)^{ur} \leq c',$$

and we arrive at (5.19) as desired.

Step 4. It remains to check (5.13). First let $r < \infty$; we apply (5.11) and make use of Hardy's inequality once more,

$$
\int_0^\varepsilon \left[\frac{\omega(f,t)}{t^\varkappa} \right]^r \frac{dt}{t} \leq c_1 \int_0^\varepsilon \left[t^{-\varkappa} \int_0^t |\nabla f|^* (\sigma^n) d\sigma \right]^r \frac{dt}{t}
$$

$$
\leq c_2 \int_0^\varepsilon \left[t^{1-\varkappa} |\nabla f|^* (t^n) \right]^r \frac{dt}{t}
$$

$$
\leq c_3 \int_0^\varepsilon \left[\tau^{\frac{1}{n}(1-\varkappa)} |\nabla f|^* (\tau) \right]^r \frac{d\tau}{\tau} \ .
$$

Finally, when $r = \infty$, we can estimate in a similar way for $0 < t < \varepsilon$, and find

$$
\frac{\omega(f,t)}{t^\varkappa} \leq c\, t^{-\varkappa} \int_0^{t^n} s^{\frac{1}{n}-1} |\nabla f|^* (s) ds
$$

$$
\leq c\, t^{-\varkappa} \left(\sup_{0<s<t^n} s^{\frac{1}{n}(1-\varkappa)} |\nabla f|^* (s) \right) \int_0^{t^n} s^{\frac{\varkappa}{n}-1} ds
$$

$$
\leq c' \sup_{0<s<\varepsilon} s^{\frac{1}{n}(1-\varkappa)} |\nabla f|^* (s).
$$

$\qquad\qquad\qquad\qquad\qquad\qquad\qquad\qquad\qquad\qquad\qquad\qquad\qquad\qquad$ $\square$

Remark 5.11 Note that Triebel obtained in [Tri01, Prop. 12.16] assertion (5.12), too, but based on a different estimate replacing (5.11) by

$$
\frac{\omega(f,t)}{t} \leq c\, |\nabla f|^{**} \left(t^{2n-1} \right) + 3 \sup_{0<\tau\leq t^2} \tau^{-\frac{1}{2}} \omega(f,\tau) \tag{5.20}
$$

for small $\varepsilon > 0$, all $t \in (0,\varepsilon)$ and all $f \in C^1$, cf. [Tri01, Prop. 12.16]. We discuss these results in Section 11.2 below; in particular, the above results also imply estimates for corresponding envelope functions, see Corollary 11.11 below. The exponent $2n-1$ (instead of n) in the first term on the right-hand side of (5.20) prevented a result like (5.13) in that case, in contrast to (5.12) where the log-term takes no notice of exponents. Neves derived an assertion similar to (5.13) from (5.20), see [Nev01b, Prop. 4.2.28]. Let us also mention that (5.12) can be derived from (5.18) directly using an extended version of Hardy's inequality obtained by Bennett and Rudnick in [BR80, Thm. 6.4].

5.3   Examples: Lipschitz spaces, Sobolev spaces

All function spaces are defined on $\mathbb{R}^n$ unless otherwise stated. In analogy
to Section 3.2 we start with some elementary examples.

Proposition 5.12 *Let* $0 < a \leq 1$. *Then*

$$\mathcal{E}_{\mathsf{C}}^{\mathrm{Lip}^a}(t) \sim t^{-(1-a)}, \quad 0 < t < 1. \tag{5.21}$$

Proof: <u>Step 1.</u>   Let $0 < a \leq 1$, $f \in \mathrm{Lip}^a$ with $\|f|\mathrm{Lip}^a\| \leq 1$. Then by
(2.25) we conclude that $\omega(f,t) \leq t^a$ for all $0 < t < 1$. Thus

$$\frac{\omega(f,t)}{t} \leq t^{-(1-a)}$$

for all $0 < t < 1$ and all $f \in \mathrm{Lip}^a$, $\|f|\mathrm{Lip}^a\| \leq 1$. This implies $\mathcal{E}_{\mathsf{C}}^{\mathrm{Lip}^a}(t) \leq t^{-(1-a)}$.

<u>Step 2.</u>   It remains to show the converse inequality. Let f_1 be a continuous
function on $\mathbb{R}^n$ which can be described as

$$f_1(x) = |x|^a \chi_{[0,1)}(|x|) + \psi(x)\chi_{[1,\infty)}(|x|), \quad x \in \mathbb{R}^n, \tag{5.22}$$

where the continuous, non-negative, monotonically decreasing function ψ may
be chosen so that f_1 is continuous, $\|f_1|C\| \sim 1$, and

$$\sup_{x \in \mathbb{R}^n} |\Delta_h f_1(x)| \sim \begin{cases} |h|^a, & |h| \leq 1, \\ 1, & |h| \geq 1. \end{cases}$$

Then $\omega(f_1,t) \sim [\min(t,1)]^a = t^a$ for $0 < t < 1$, and $\|f_1|\mathrm{Lip}^a\| \sim 1$.
Consequently, for any $0 < t < 1$,

$$\mathcal{E}_{\mathsf{C}}^{\mathrm{Lip}^a}(t) \geq \frac{\omega(f_1,t)}{t} \sim \frac{t^a}{t} = t^{-(1-a)}.$$

This proves (5.21).   □

Proposition 5.13 *Let* $0 < q \leq \infty$, $\alpha > \frac{1}{q}$ *(with* $\alpha \geq 0$ *if* $q = \infty$*). Then*

$$\mathcal{E}_{\mathsf{C}}^{\mathrm{Lip}^{(1,-\alpha)}_{\infty,q}}(t) \sim |\log t|^{\alpha - \frac{1}{q}}, \quad 0 < t < \frac{1}{2}. \tag{5.23}$$

P r o o f :   Step 1.   First let $q = \infty$; then the corresponding upper bound in (5.23) follows from (2.35) by straightforward calculation. When $q < \infty$, we make use of the monotonicity of $\frac{w(f,t)}{t}$ (up to constants) and conclude that for any τ, $0 < \tau < \frac{1}{2}$, and any $f \in \mathrm{Lip}^{(1,-\alpha)}_{\infty,q}$, $\|f|\mathrm{Lip}^{(1,-\alpha)}_{\infty,q}\| \leq 1$,

$$
1 \geq \left(\int_0^{\frac{1}{2}} \left[\frac{w(f,t)}{t\,|\log t|^\alpha} \right]^q \frac{dt}{t} \right)^{1/q} \geq \left(\int_0^\tau \left[\frac{w(f,t)}{t\,|\log t|^\alpha} \right]^q \frac{dt}{t} \right)^{1/q}
$$

$$
\geq c\,\frac{w(f,\tau)}{\tau} \left(\int_0^\tau \frac{dt}{t\,|\log t|^{\alpha q}} \right)^{1/q} \geq c'\,\frac{w(f,\tau)}{\tau}\,|\log \tau|^{-(\alpha-\frac{1}{q})}.
$$

We benefit from $\alpha q > 1$. Hence there is some $c > 0$ such that for any $\tau > 0$ and any $f \in \mathrm{Lip}^{(1,-\alpha)}_{\infty,q}$, $\|f|\mathrm{Lip}^{(1,-\alpha)}_{\infty,q}\| \leq 1$,

$$
\frac{w(f,\tau)}{\tau} \leq c\,|\log \tau|^{\alpha-\frac{1}{q}},
$$

yielding the upper estimate in (5.23) for $q < \infty$.

Step 2.   We verify the estimate from below in (5.23). We slightly modify the function f_1 from (5.22) and adapt it to our setting: for $0 < s < \frac{1}{2}$, $0 < q \leq \infty$ and $\alpha > \frac{1}{q}$ let

$$
f_s(x) = |\log s|^{\alpha-\frac{1}{q}}\,|x|\,\chi_{[0,s)}(|x|) + \psi(x)\chi_{[s,\infty)}(|x|), \quad x \in \mathbb{R}^n,
$$

and ψ is chosen as above. Then $w(f_s,t) \sim \min(t,s)\,|\log s|^{\alpha-\frac{1}{q}}$. Thus $\|f|C\| \sim s\,|\log s|^{\alpha-\frac{1}{q}} \leq c_{\alpha,q}$ for $0 < s < \frac{1}{2}$. On the other hand,

$$
\left(\int_0^{\frac{1}{2}} \left[\frac{w(f_s,t)}{t\,|\log t|^\alpha} \right]^q \frac{dt}{t} \right)^{1/q} \leq c\,|\log s|^{\alpha-\frac{1}{q}} \left(\int_0^{\frac{1}{2}} \left[\frac{\min(s,t)}{t\,|\log t|^\alpha} \right]^q \frac{dt}{t} \right)^{1/q}
$$

$$
\leq c_1\,|\log s|^{\alpha-\frac{1}{q}} \left(\int_0^s \left[\frac{t}{t\,|\log t|^\alpha} \right]^q \frac{dt}{t} \right)^{1/q}
$$

$$
+ c_2\,s\,|\log s|^{\alpha-\frac{1}{q}} \left(\int_s^{\frac{1}{2}} \left[\frac{1}{t\,|\log t|^\alpha} \right]^q \frac{dt}{t} \right)^{1/q}
$$

$$
\leq c_3\,|\log s|^{\alpha-\frac{1}{q}} \left(\int_0^s \frac{dt}{t\,|\log t|^{\alpha q}} \right)^{1/q} + c_4
$$

$$
\leq c_5\,|\log s|^{\alpha-\frac{1}{q}}\,|\log s|^{-(\alpha-\frac{1}{q})} + c_4 \leq C, \qquad (5.24)
$$

(with obvious modifications when $q = \infty$) so that $f_s \in \mathrm{Lip}^{(1,-\alpha)}_{\infty,q}$ and $\|f_s|\mathrm{Lip}^{(1,-\alpha)}_{\infty,q}\| \sim 1$. On the other hand,

$$\mathcal{E}^{\mathrm{Lip}^{(1,-\alpha)}_{\infty,q}}_{C}(t) \geq \sup_{f_s} \frac{\omega(f_s,t)}{t} \sim \sup_{s>0} |\log s|^{\alpha-\frac{1}{q}} \frac{\min(t,s)}{t} \geq \sup_{s>t} |\log s|^{\alpha-\frac{1}{q}}$$

$$\sim |\log t|^{\alpha-\frac{1}{q}},$$

where the last estimate results from $0 < t < s < \frac{1}{2}$. $\qquad\square$

Proposition 5.14 *Let* $0 < a < 1$, $0 < q \leq \infty$, *and* $\alpha \in \mathbb{R}$. *Then*

$$\mathcal{E}^{\mathrm{Lip}^{(a,-\alpha)}_{\infty,q}}_{C}(t) \sim t^{-(1-a)} |\log t|^{\alpha}, \quad 0 < t < \frac{1}{2}.$$

P r o o f : The proof is simply a combination of those for Propositions 5.12 and 5.13, where the extremal functions f_s can now be chosen as

$$f_s(x) = |\log s|^{\alpha} s^{a-1} |x| \chi_{[0,s)}(|x|) + \psi(x)\chi_{[s,\infty)}(|x|), \quad x \in \mathbb{R}^n,$$

$s > 0$ small; the modifications are clear otherwise. $\qquad\square$

Having dealt in Proposition 5.13 with the modifications $\mathrm{Lip}^{(1,-\alpha)}_{\infty,q}$ of the "upper borderline case" Lip^1, we finally concentrate on the basic space C.

Proposition 5.15 *Let* C *be as above. Then*

$$\mathcal{E}^{C}_{C}(t) \sim t^{-1}, \quad 0 < t < 1. \tag{5.25}$$

P r o o f : The upper estimate simply comes from $\omega(f,t) \leq 2\,\|f|C\|$ for all $t > 0$ and all $f \in C$, see Remark 5.6. Concerning the lower bound we use functions $f_n(x)$, $n \in \mathbb{N}$, defined by

$$f_n(x) = \begin{cases} 1 - n|x|, & |x| \leq \frac{1}{n}, \\ 0 & , \text{ otherwise.} \end{cases} \tag{5.26}$$

Clearly, $\|f_n|C\| = 1$ and $\omega(f_n,t) = 1$ for $t \geq \frac{1}{n}$. Thus choose for $t_0 > 0$ some $m \in \mathbb{N}$ with $m \geq \frac{1}{t_0}$; then

$$\mathcal{E}^{C}_{C}(t_0) = \sup_{\|f|C\|\leq 1} \frac{\omega(f,t_0)}{t_0} \geq \frac{\omega(f_m,t_0)}{t_0} = \frac{1}{t_0},$$

completing the proof. $\qquad\square$

In view of Theorem 2.28(i) we have $W_p^k \hookrightarrow C$ for $k > \frac{n}{p}$, $1 \le p < \infty$, or $k = n$ and $p = 1$, so that the concept of continuity envelopes makes sense. On the other hand, by (2.39), (2.43), combined with Proposition 5.3(ii) we conclude that unbounded continuity envelope functions appear in the following settings:

$$W_p^k \not\hookrightarrow \mathrm{Lip}^1 \quad \text{if} \quad \left\{ \begin{array}{c} \frac{n}{p} < k < \frac{n}{p} + 1,\ 1 \le p < \infty \\ k = \frac{n}{p} + 1,\ 1 < p < \infty \\ k = n \quad,\quad p = 1 \end{array} \right\} \tag{5.27}$$

We start with the super-critical strip $\frac{n}{p} < k < \frac{n}{p} + 1$, $1 \le p < \infty$.

Proposition 5.16 Let $1 \le p < \infty$, $k \in \mathbb{N}$, with $\frac{n}{p} < k < \frac{n}{p} + 1$. Then

$$\mathcal{E}_C^{W_p^k}(t) \sim t^{-\left(\frac{n}{p}+1-k\right)}, \quad 0 < t < 1. \tag{5.28}$$

P r o o f : In view of Theorem 2.28(i), in particular, (2.41),

$$W_p^k \hookrightarrow \mathrm{Lip}^a, \quad a = k - \frac{n}{p}, \tag{5.29}$$

see also [Zie89, Thm. 2.4.4]. Thus Propositions 5.3(iii) and 5.12 imply

$$\mathcal{E}_C^{W_p^k}(t) \le c\, t^{-\left(\frac{n}{p}+1-k\right)}, \quad 0 < t < 1.$$

Conversely, we return to our examples already used in (3.35),

$$f_R(x) = R^{k-\frac{n}{p}}\, \psi\left(R^{-1}x\right), \quad x \in \mathbb{R}^n, \quad 0 < R < 1,$$

where $\psi(x)$ is given by (3.36). Recall that with $g_R := \|\psi|W_p^k\|^{-1}f_R \in W_p^k$ we obtain

$$\left\|g_R|W_p^k\right\| = \frac{\|f_R|W_p^k\|}{\|\psi|W_p^k\|} \le 1, \quad 0 < R < 1.$$

On the other hand, by construction,

$$\omega(g_R, R) = \frac{R^{k-\frac{n}{p}}}{\|\psi|W_p^k\|\, e} \sim R^{k-\frac{n}{p}}, \quad 0 < R < 1,$$

that is,

$$\mathcal{E}_C^{W_p^k}(t) \ge \sup_{0 < R < 1} \frac{\omega(g_R, t)}{t} \ge \frac{\omega(g_t, t)}{t} \ge c\, t^{k-\frac{n}{p}-1},$$

for $0 < t < 1$. $\square$

We consider the limiting case $k = \frac{n}{p} + 1$, $1 < p < \infty$, now.

Proposition 5.17 *Let* $1 < p < \infty$ *be such that* $\frac{n}{p} + 1 = k \in \mathbb{N}$. *Then*

$$\mathcal{E}_{\mathsf{C}}^{W_p^{1+n/p}}(t) \sim |\log t|^{\frac{1}{p'}}, \quad 0 < t < \frac{1}{2}. \tag{5.30}$$

Proof: We make use of our earlier results on growth envelopes, in particular, the result by Hansson [Han79, (3.13)] recalled in (4.23), and a more general statement in Proposition 5.10(ii). Obviously, by density arguments it is sufficient to prove (5.30) for all sufficiently smooth functions. On the other hand, $f \in W_p^{1+n/p}$ implies $|\nabla f| \in W_p^{n/p}$, and we may apply Proposition 4.11, in particular, (4.23) to $|\nabla f|$. Together with (5.12) with $1 < r = p < \infty$, $u = 1 > \frac{1}{p'}$, and $0 < \varepsilon < 1$ this yields the existence of $c > 0$ such that for all $f \in C^1$,

$$\left(\int_0^\varepsilon \left[\frac{\omega(f,t)}{t \, |\log t|} \right]^p \frac{dt}{t} \right)^{1/p} \leq c \left(\int_0^\varepsilon \left[\frac{|\nabla f|^*(t)}{|\log t|} \right]^p \frac{dt}{t} \right)^{1/p}$$

$$\leq c' \left\| |\nabla f| \, |W_p^{n/p} \right\|$$

$$\leq c'' \left\| f | W_p^{1+n/p} \right\|. \tag{5.31}$$

On the other hand, for arbitrary τ, $0 < \tau < \varepsilon$, the left-hand side of (5.31) can be estimated from below by

$$\left(\int_0^\tau \left[\frac{\omega(f,t)}{t \, |\log t|} \right]^p \frac{dt}{t} \right)^{1/p} \geq c \, \frac{\omega(f,\tau)}{\tau} \left(\int_0^\tau \frac{dt}{t \, |\log t|^p} \right)^{1/p}$$

$$\geq c' \, \frac{\omega(f,\tau)}{\tau} \, |\log \tau|^{-\frac{1}{p'}}$$

due to the monotonicity of $\frac{\omega(f,t)}{t}$, see (5.4) or [DL93, Ch. 2, Lemma 6.1]. Thus we have by (5.31) for all $f \in C^1 \cap W_p^{1+n/p}$ with $\left\| f | W_p^{1+n/p} \right\| \leq 1$ and all τ, $0 < \tau < \varepsilon$,

$$\frac{\omega(f,\tau)}{\tau} \leq c \, |\log \tau|^{\frac{1}{p'}},$$

i.e.,

$$\mathcal{E}_{\mathsf{C}}^{W_p^{1+n/p}}(t) \leq c \, |\log t|^{\frac{1}{p'}}, \quad 0 < t < \varepsilon.$$

For the converse we modify an argument from the proof of Proposition 3.27. In particular, we adapt the construction (3.44) by

$$h_m(x) = \sum_{j=0}^{m-1} 2^{-j} \, \psi\left(2^j x\right), \quad m \in \mathbb{N}, \tag{5.32}$$

and ψ given by (3.36). Then we obtain in the same way as described above,

$$\left\| h_m | W_p^{1+n/p} \right\| \le c' \, m^{1/p} \left\| \psi | W_p^{1+n/p} \right\| \le C \, m^{1/p}, \quad m \in \mathbb{N}. \tag{5.33}$$

On the other hand, we have for $k = 1, \dots, m$,

$$\omega\left(h_m, 2^{-k}\right) \ge \sum_{j=0}^{m-1} 2^{-j} \psi(0) - \sum_{j=0}^{k-1} 2^{-j} \psi\left(2^{j-k}\theta\right) \ge c \sum_{j=0}^{k-1} 2^{-j+j-k}$$

$$\ge c\,k\,2^{-k} \tag{5.34}$$

where $\theta \in \mathbb{C}^n$, $|\theta| = 1$. Let t, $0 < t < \frac{1}{2}$, be given and choose $m_0 \in \mathbb{N}$ such that $2^{-(m_0+1)} < t \le 2^{-m_0}$, i.e., $m_0 \sim |\log t|$. Then by (5.33), (5.34) with $k = m_0$,

$$\mathcal{E}_C^{W_p^{1+n/p}}(t) \ge c \sup_{m \in \mathbb{N}} m^{-\frac{1}{p}} \frac{\omega(h_m, t)}{t} \ge c\,m_0^{-\frac{1}{p}} \frac{\omega(h_{m_0}, t)}{t}$$

$$\ge c'\,m_0^{-\frac{1}{p}} \frac{\omega(h_{m_0}, 2^{-m_0})}{2^{-m_0}} \ge c''\,m_0^{-\frac{1}{p}+1} \ge C\,|\log t|^{-\frac{1}{p}+1}.$$

This completes the proof. $\square$

Remark 5.18 Note that one part of the estimate (5.30) could have been obtained by (2.60) in connection with Propositions 5.3(iii) and 5.13. This would, however, not cover the proof for the fine index $u_C^{W_p^{1+n/p}}$, see Proposition 6.8 below. Therefore we stressed an alternative line of argument. In Part II, more precisely, in Theorem 9.4(i) we obtain the counterpart of (5.30) in a more general setting.

It remains to deal with the case $k = n$, $p = 1$.

Proposition 5.19 *We have*

$$\mathcal{E}_C^{W_1^n}(t) \sim t^{-1}, \quad 0 < t < \frac{1}{2}. \tag{5.35}$$

P r o o f :   Clearly, Proposition 5.15 together with the embedding (2.38) and Proposition 5.3(iii) imply the corresponding upper estimate. Conversely, note that the construction from Proposition 5.16 remains valid in case of $k = n$, $p = 1$, thus leading for

$$g_R(x) = \frac{\psi\left(R^{-1}x\right)}{\|\psi | W_1^n\|}, \quad 0 < R < 1,$$

to

$$\|g_R|W_1^n\| \leq 1, \qquad \omega\left(g_R, \frac{R}{2}\right) \sim 1, \quad 0 < R < 1,$$

that is,

$$\mathcal{E}_{\mathsf{C}}^{W_1^n}(t) \geq \sup_{0<R<1} \frac{\omega(g_R,t)}{t} \geq \frac{\omega(g_{2t},t)}{t} \geq c\,t^{-1}$$

for $0 < t < \dfrac{1}{2}$. $\qquad\qquad\qquad\qquad\qquad\qquad\qquad\qquad\qquad$ □

Chapter 6

Continuity envelopes $\mathfrak{E}_\mathsf{C}$

We do not want to repeat our introductory remarks as presented in Chapter 4. Obviously the continuity envelope function as given in Chapter 5 is a suitable counterpart of the growth envelope function handled in Chapter 3; so a parallel argument may serve as motivation when complementing $\mathcal{E}_\mathsf{C}^X$ by the index u_C^X, too.

6.1 Definition

Analogous to the situation described at the beginning of Section 4.1 we shall introduce the Borel measure μ_C associated with the function ψ as described in Section 4.1, and equivalent to $1/\mathcal{E}_\mathsf{C}^X$ for some function space X with (5.8) and $X \not\hookrightarrow \mathrm{Lip}^1$, $\psi(t) \sim 1/\mathcal{E}_\mathsf{C}^X(t)$, $0 < t < \varepsilon$. Then (granted that $\mathcal{E}_\mathsf{C}^X$ was continuously differentiable) we obtain

$$\mu_\mathsf{C}(\mathrm{d}t) \sim -\frac{\left(\mathcal{E}_\mathsf{C}^X\right)'(t)}{\mathcal{E}_\mathsf{C}^X(t)}\, \mathrm{d}t \tag{6.1}$$

for small $t > 0$.

Definition 6.1 *Let* $X \hookrightarrow C$ *be a function space on* $\mathbb{R}^n$ *with* (5.8), $X \not\hookrightarrow \mathrm{Lip}^1$ *and continuity envelope function* $\mathcal{E}_\mathsf{C}^X$. *Assume* $\varepsilon > 0$. *The index* u_C^X, $0 < u_\mathsf{C}^X \le \infty$, *is defined as the infimum of all numbers* v, $0 < v \le \infty$, *such that*

$$\left(\int\limits_0^\varepsilon \left[\frac{\omega(f,t)}{t\, \mathcal{E}_\mathsf{C}^X(t)} \right]^v \mu_\mathsf{C}(\mathrm{d}t) \right)^{1/v} \le c\, \|f|X\| \tag{6.2}$$

(with the usual modification if $v = \infty$*) holds for some* $c > 0$ *and all* $f \in X$. *Then*

$$\mathfrak{E}_\mathsf{C}(X) = \left(\mathcal{E}_\mathsf{C}^X(\cdot), u_\mathsf{C}^X \right) \tag{6.3}$$

is called the continuity envelope for the function space X.

Remark 6.2 Proposition 5.3(iv) (with $\varkappa \equiv 1$) implies that (6.2) holds with $v = \infty$ in any case; but – depending upon the underlying function space X – there might be some smaller v such that (6.2) is still satisfied. As Proposition 4.1(i) can be applied to the above case, that is, $\psi \sim 1/\mathcal{E}_C$ and $g(t) \sim \frac{\omega(f,t)}{t}$, without any difficulties, we have the monotonicity of (6.2) in v.

The question posed in Section 4.1, that is, under which assumptions

$$u_C^X = \inf \{v : 0 < v \leq \infty, \, v \quad \text{satisfies} \quad (6.2)\} \tag{6.4}$$

is in fact a *minimum*, makes sense in that context, too, but is likewise open in general. Again, all the examples studied below are such (possibly special) cases where u_C^X satisfies (6.2).

Remark 6.3 In analogy to Remark 4.4 we handle the case when $X \hookrightarrow \mathrm{Lip}^1$ separately. Parallel to Remark 4.4 we can include this situation by putting $u_C^X := \infty$ as for bounded $\mathcal{E}_C^X$, that is, by Proposition 5.3(ii), when $X \hookrightarrow \mathrm{Lip}^1$, (6.2) can be replaced by

$$\sup_{0 < t < \varepsilon} \frac{\omega(f,t)}{t} \leq c \, \|f|X\|,$$

for some $c > 0$ and all $f \in X$.

We give the counterpart of Proposition 4.5 in terms of continuity envelopes.

Proposition 6.4 *Let* $X_i \hookrightarrow C$, $i = 1, 2$, *be some function spaces on* $\mathbb{R}^n$ *with* $X_1 \hookrightarrow X_2$. *Assume for their continuity envelope functions*

$$\mathcal{E}_C^{X_1}(t) \sim \mathcal{E}_C^{X_2}(t), \quad 0 < t < \varepsilon. \tag{6.5}$$

Then for the corresponding indices $u_C^{X_i}$, $i = 1, 2$, *we have*

$$u_C^{X_1} \leq u_C^{X_2}. \tag{6.6}$$

P r o o f :   The proof copies that of Proposition 4.5.   □

Analogous to Remark 4.6 we could *interpret* the meaning of (6.2) in terms of sharp embeddings in, say, target spaces of type $\mathrm{Lip}_{\infty,q}^{(1,-\alpha)}$. We shall not perform this here in detail.

Remark 6.5 In analogy to Remark 5.11 let us mention that Proposition 5.10 leads to an assertion about related fine indices, see Corollary 11.11 below.

6.2   Examples: Lipschitz spaces, Sobolev spaces

We return to our examples from Section 5.3.

Theorem 6.6

(i)   *Let $0 < a \leq 1$. Then*

$$\mathfrak{E}_C\left(\mathrm{Lip}^a\right) = \left(t^{-(1-a)},\ \infty\right). \tag{6.7}$$

(ii)   *Let $0 < q \leq \infty$, $\alpha > \frac{1}{q}$ (with $\alpha \geq 0$ if $q = \infty$). Then*

$$\mathfrak{E}_C\left(\mathrm{Lip}^{(1,-\alpha)}_{\infty,q}\right) = \left(|\log t|^{\alpha - \frac{1}{q}},\ q\right). \tag{6.8}$$

(iii)   *Let $0 < a < 1$, $0 < q \leq \infty$, and $\alpha \in \mathbb{R}$. Then*

$$\mathfrak{E}_C\left(\mathrm{Lip}^{(a,-\alpha)}_{\infty,q}\right) = \left(t^{-(1-a)}|\log t|^{\alpha},\ q\right). \tag{6.9}$$

(iv)   *We have*

$$\mathfrak{E}_C(\,C\,) = \left(t^{-1},\ \infty\right). \tag{6.10}$$

P r o o f : Step 1. We prove (i). Let first $a < 1$. Note that Proposition 5.12 yields $\mathcal{E}_C^{\mathrm{Lip}^a}(t) \sim t^{-(1-a)}$, $0 < t < 1$. Thus, taking (6.1) and (6.2) into account, we have to check for which numbers v there exists $c > 0$ such that for all $f \in \mathrm{Lip}^a$,

$$\left(\int_0^\varepsilon \left[\frac{\omega(f,t)}{t^a}\right]^v \frac{dt}{t}\right)^{1/v} \leq c\,\|f|\mathrm{Lip}^a\|.$$

In view of (2.25) this leads to $v = \infty$ only. Finally, the case $a = 1$ is covered by Remark 6.3.

Step 2. We come to (ii). In Proposition 5.13 we obtained $\mathcal{E}_C^{\mathrm{Lip}^{(1,-\alpha)}_{\infty,q}}(t) \sim |\log t|^{\alpha - \frac{1}{q}}$, $0 < t < \frac{1}{2}$. The corresponding index $u_C^{\mathrm{Lip}^{(1,-\alpha)}_{\infty,q}}$ in (6.8) is the least number v such that there is a positive number c with

$$\left(\int_0^\varepsilon \left[\frac{\omega(f,t)}{t\,|\log t|^{\alpha - 1/q + 1/v}}\right]^v \frac{dt}{t}\right)^{1/v} \leq c\,\left\|f|\mathrm{Lip}^{(1,-\alpha)}_{\infty,q}\right\|$$

for all $f \in \text{Lip}_{\infty, q}^{(1, -\alpha)}$, recall (6.1). In view of (2.35), for which numbers v do we obtain (at least locally)

$$\text{Lip}_{\infty, q}^{(1, -\alpha)} \quad \hookrightarrow \quad \text{Lip}_{\infty, v}^{(1, -(\alpha - \frac{1}{q} + \frac{1}{v}))}$$

we are led to ask. We apply Proposition 2.23 with $p = \infty$, $r = v$, and $\beta = \alpha - \frac{1}{q} + \frac{1}{v}$ to obtain $v \geq q$.

Step 3.   We verify (iii). By Proposition 5.14 we look for the smallest number v such that

$$\left(\int_0^\varepsilon \left[\frac{\omega(f, t)}{t^a \, |\log t|^\alpha} \right]^v \frac{dt}{t} \right)^{1/v} \leq c \, \left\| f | \text{Lip}_{\infty, q}^{(a, -\alpha)} \right\| \tag{6.11}$$

for some positive number c and all $f \in \text{Lip}_{\infty, q}^{(a, -\alpha)}$. Plainly, (2.35) yields $u_C^{\text{Lip}_{\infty, q}^{(a, -\alpha)}} \leq q$. In order to show the converse inequality we proceed by contradiction and modify our argument presented in [Har00b, Prop. 16] concerning $\text{Lip}_{\infty, q}^{(1, -\alpha)}$ slightly. Let $f_\varkappa \in C$ be such that

$$\omega(f_\varkappa, t) \sim t^a \, |\log t|^\varkappa$$

for small $t > 0$. Clearly, $f_\varkappa$ belongs to $\text{Lip}_{\infty, q}^{(a, -\alpha)}$ if, and only if, $\varkappa < \alpha - \frac{1}{q}$. Assume now that (6.11) holds for some number $v < q$ and all $f \in \text{Lip}_{\infty, q}^{(a, -\alpha)}$. Then we can choose $\varkappa$ such that $\alpha - \frac{1}{v} \leq \varkappa < \alpha - \frac{1}{q}$ and arrive at a contradiction in (6.11) immediately.

Step 4.   It remains to show (iv). In view of Remark 6.2 it is sufficient to prove $u_C^C \geq \infty$. We proceed by contradiction, i.e., we disprove that there is a number $v < \infty$ such that

$$\left(\int_0^\varepsilon [\omega(f, t)]^v \frac{dt}{t} \right)^{1/v} \leq c \, \|f|C\| \tag{6.12}$$

holds for all $f \in C$, $\|f|C\| \leq 1$. We make use of the functions f_n given by (5.26), $n \in \mathbb{N}$. Recall $\|f_n|C\| = 1$ and $\omega(f_n, t) = 1$ for $t \geq \frac{1}{n}$, then (6.12) implies that there exists $c > 0$ such that for all $n \in \mathbb{N}$

$$1 = \|f_n|C\| \geq \frac{1}{c} \left(\int_0^\varepsilon [\omega(f_n, t)]^v \frac{dt}{t} \right)^{1/v}$$

$$\geq \frac{1}{c} \left(\int_{\frac{1}{n}}^\varepsilon [\omega(f_n, t)]^v \frac{dt}{t} \right)^{1/v} \geq c' \, |\log n|^{\frac{1}{v}} \, .$$

However, this is impossible for all $n \in \mathbb{N}$, and so $u_C^C = \infty$.   □

Proposition 6.7 *Let* $1 \le p < \infty$, $k \in \mathbb{N}$, *with* $\frac{n}{p} < k < \frac{n}{p} + 1$. *Then*

$$\mathfrak{E}_C\left(W_p^k\right) = \left(t^{-\left(\frac{n}{p}+1-k\right)}, p\right). \tag{6.13}$$

P r o o f :   In view of Proposition 5.16 and Definition 6.1 it remains to prove that

$$\left(\int_0^\varepsilon \left[\frac{\omega(f,t)}{t^{k-\frac{n}{p}}}\right]^v \frac{dt}{t}\right)^{1/v} \le c\, \|f|W_p^k\| \tag{6.14}$$

holds for some $c > 0$ and all $f \in W_p^k$ if, and only if, $v \ge p$. First we show that there cannot be a number $v < p$ that guarantees (6.14) for all $f \in W_p^k$; this is done by contradiction using a refined construction of "extremal" functions based upon (3.35) with $R = 2^{-j}$. Let $b = \{b_j\}_{j \in \mathbb{N}}$ be a sequence of non-negative numbers, and let $x^0 \in \mathbb{R}^n$, $|x^0| > 4$, such that supp $\psi\left(2^j \cdot -x^0\right) \cap$ supp $\psi\left(2^r \cdot -x^0\right) = \emptyset$ for $j \ne r$, $j, r \in \mathbb{N}_0$, and ψ given by (3.36). Then

$$f_b(x) := \sum_{j=1}^\infty 2^{-j(k-\frac{n}{p})}\, b_j\, \psi\left(2^j x - x^0\right), \quad x \in \mathbb{R}^n, \tag{6.15}$$

belongs to W_p^k for $b \in \ell_p$, and $\|f_b|W_p^k\| \le c\|b|\ell_p\|$, see (4.16). On the other hand, by construction

$$\omega\left(f_b, 2^{-j}\right) \ge c\, 2^{-j(k-\frac{n}{p})}\, b_j, \quad j \in \mathbb{N}_0.$$

For convenience we may assume $b_1 = \cdots = b_{J-1} = 0$, where J is suitably chosen such that $2^{-J} \sim \varepsilon$ given by (6.14). Then by monotonicity arguments,

$$\left(\int_0^\varepsilon \left[\frac{\omega(f_b,t)}{t^{k-\frac{n}{p}}}\right]^v \frac{dt}{t}\right)^{1/v} \sim \left(\sum_{j=J}^\infty \left[2^{-j(\frac{n}{p}-k)}\, \omega\left(f_b, 2^{-j}\right)\right]^v\right)^{1/v}$$

$$\ge c\left(\sum_{j=J}^\infty \left[2^{-j(\frac{n}{p}-k)}b_j\, 2^{-j(k-\frac{n}{p})}\right]^v\right)^{1/v} \sim \|b|\ell_v\|.$$

Together with $\|f_b|W_p^k\| \le c\|b|\ell_p\|$ our assumption (6.14) thus implies $\|b|\ell_p\| \ge c\|b|\ell_v\|$ for arbitrary sequences of non-negative numbers. This obviously requires $v \ge p$. It remains to verify the converse, i.e.,

$$\left(\int_0^\varepsilon \left[\frac{\omega(f,t)}{t^{k-\frac{n}{p}}}\right]^p \frac{dt}{t}\right)^{1/p} \le c\, \|f|W_p^k\|. \tag{6.16}$$

Here we gain from a more general estimate and our previous observations for growth envelopes. In particular, we use Proposition 5.10(iii), which gives us for $1 \le p < \infty$, $0 < \varkappa < 1$, and $0 < \varepsilon < 1$ the existence of $c > 0$ such that for all $f \in C^1$,

$$\int_0^\varepsilon \left[\frac{\omega(f,t)}{t^\varkappa}\right]^p \frac{dt}{t} \le c \int_0^\varepsilon \left[t^{\frac{1}{n}(1-\varkappa)} |\nabla f|^*(t)\right]^p \frac{dt}{t} \qquad (6.17)$$

Obviously, by density arguments it is sufficient to prove (6.16) for all sufficiently smooth functions. On the other hand, $f \in W_p^k$ with $\frac{n}{p} < k < \frac{n}{p} - 1$, implies $|\nabla f| \in W_p^{k-1}$, $0 \le k - 1 < \frac{n}{p}$, and we may apply Proposition 4.10, in particular, (4.14) to $|\nabla f|$. Together with (6.17) and $\varkappa = k - \frac{n}{p}$ this completes the proof,

$$\left(\int_0^\varepsilon \left[\frac{\omega(f,t)}{t^{k-\frac{n}{p}}}\right]^p \frac{dt}{t}\right)^{1/p} \le c \left(\int_0^\varepsilon \left[t^{\frac{1}{p}-\frac{k-1}{n}} |\nabla f|^*(t)\right]^p \frac{dt}{t}\right)^{1/p}$$

$$\le c' \||\nabla f| \, |W_p^{k-1}\| \le c'' \|f|W_p^k\|.$$

$\square$

We continue Proposition 5.17 dealing with the case $k = \frac{n}{p} + 1$.

Proposition 6.8 *Let* $1 < p < \infty$ *be such that* $\frac{n}{p} + 1 = k \in \mathbb{N}$. *Then*

$$\mathfrak{E}_C\left(W_p^{1+n/p}\right) = \left(|\log t|^{\frac{1}{p'}}, \, p\right).$$

Proof: In view of $\mathcal{E}_C^{W_p^{1+n/p}}(t) \sim |\log t|^{1/p'}$ by (5.30), and

$$\mu_C(dt) \sim \frac{1}{|\log t|} \frac{dt}{t}, \quad 0 < t < \varepsilon < \frac{1}{2},$$

we have to show that

$$\left(\int_0^\varepsilon \left[\frac{\omega(f,t)}{t \, |\log t|^{1/p'+1/v}}\right]^v \frac{dt}{t}\right)^{1/v} \le c \, \|f|W_p^{1+n/p}\| \qquad (6.18)$$

holds for all $f \in W_p^{1+n/p}$ if, and only if, $v \ge p$, i.e., $u_C^{W_p^{1+n/p}} = p$. Obviously (5.31) gives $u_C^{W_p^{1+n/p}} \le p$ and it suffices to prove that (6.18) fails for $v < p$. We stress an argument parallel to the proof of Proposition 4.11 and proceed by contradiction. Assume that (6.18) holds for some $v < p$ and consider slightly modified functions of type (5.32), i.e.,

$$h_{m,b}(x) = \sum_{j=0}^{m-1} 2^{-j}\, b_j\, \psi\left(2^j x\right), \quad m \in \mathbb{N}, \tag{6.19}$$

where ψ is given by (3.36) and $b_j \geq 0$, $j = 0, \ldots, m-1$. Then just as in (5.33),

$$\left\|h_{m,b}|W_p^{1+n/p}\right\| \leq \sum_{|\alpha|\leq 1+\frac{n}{p}} \|D^\alpha h_{m,b}|L_p\| \leq c\left\|\psi|W_p^{1+n/p}\right\| \left(\sum_{j=0}^{m-1} b_j^p\right)^{1/p}$$

$$\leq c'\left(\sum_{j=0}^{m-1} b_j^p\right)^{1/p}.$$

If we choose

$$b_j = (j+1)^{-\frac{1}{p}}\left(1 + \log(j+1)\right)^{-\frac{1}{v}}, \quad j = 0, \ldots, m-1,$$

then

$$\left\|h_{m,b}|W_p^{1+n/p}\right\| \leq c\left(\sum_{j=1}^{m} \frac{1}{j\left(1 + \log j\right)^{p/v}}\right)^{1/p} \leq c' \tag{6.20}$$

uniformly in $m \in \mathbb{N}$ because $p > v$. On the other hand, parallel to (5.34) one obtains for $r = 1, \ldots, m$,

$$\omega\left(h_{m,b}, 2^{-r}\right) \geq c\, 2^{-r} \sum_{j=0}^{r-1} b_j \geq c\, 2^{-r}\, r\, b_{r-1} = c\, 2^{-r}\, r^{\frac{1}{p'}}\left(1 + \log r\right)^{-\frac{1}{v}}, \tag{6.21}$$

so that finally (6.18) implies for sufficiently large $m \in \mathbb{N}$,

$$\left\|h_{m,b}|W_p^{1+n/p}\right\| \geq c_1 \left(\int_0^\varepsilon \left[\frac{\omega(h_{m,b}, t)}{t\,|\log t|^{1/p'+1/v}}\right]^v \frac{dt}{t}\right)^{1/v}$$

$$\geq c_2 \left(\sum_{r=1}^{m} \left[\frac{\omega(h_{m,b}, 2^{-r})}{2^{-r}\, r^{1/p'+1/v}}\right]^v\right)^{1/v}$$

$$\geq c_3 \left(\sum_{r=1}^{m} \frac{1}{r\left(1 + \log r\right)}\right)^{1/v}$$

which does not converge for $m \to \infty$ in contrast to the left-hand side. Hence by this contradiction we conclude $v \geq p$ and the proof is complete. $\square$

Remark 6.9 The above result reappears as a special case of Theorem 9.4(i) below.

We finally return to the case $k = n$, $p = 1$, but have not yet a complete result.

Lemma 6.10 *We have*

$$\mathcal{E}_{\mathsf{C}}^{W_1^n}(t) \sim t^{-1}, \quad 0 < t < \varepsilon, \qquad and \qquad u_{\mathsf{C}}^{W_1^n} \geq 1. \qquad (6.22)$$

P r o o f : In view of Proposition 5.19 we have to show the existence of some $c > 0$ such that for all $f \in W_1^n$

$$\left(\int_0^\varepsilon [\omega(f,t)]^v \, \frac{dt}{t} \right)^{1/v} \leq c \, \|f|W_1^n\| \qquad (6.23)$$

implies $v \geq 1$, i.e., $u_{\mathsf{C}}^{W_1^n} \geq 1$. Plainly, a slightly adapted version of the argument used in the proof of Proposition 6.8 immediately covers this case. $\square$

Part II

Results in function spaces, and applications

Chapter 7

Function spaces and embeddings

Function spaces of Besov or Triebel-Lizorkin type, $B_{p,q}^s$ and $F_{p,q}^s$, respectively, will serve as outstanding examples in the sequel. Thus we recall briefly the basic ingredients needed for their introduction. In addition, we collect well-known characterisations and embedding results. Note that all spaces are defined on $\mathbb{R}^n$ unless otherwise stated; so we shall omit the "$\mathbb{R}^n$" in the sequel.

7.1 Spaces of type $B_{p,q}^s$, $F_{p,q}^s$

The Schwartz space $\mathcal{S}$ and its dual $\mathcal{S}'$ of all complex-valued tempered distributions have their usual meaning here. Let $\varphi_0 = \varphi \in \mathcal{S}$ be such that

$$\operatorname{supp} \varphi \subset \{y \in \mathbb{R}^n : |y| < 2\} \quad \text{and} \quad \varphi(x) = 1 \quad \text{if} \quad |x| \le 1, \tag{7.1}$$

and for each $j \in \mathbb{N}$ let $\varphi_j(x) = \varphi(2^{-j}x) - \varphi(2^{-j+1}x)$. Then $\{\varphi_j\}_{j=0}^{\infty}$ forms a *smooth dyadic resolution of unity*. Given any $f \in \mathcal{S}'$, we denote by $\mathcal{F}f$ and $\mathcal{F}^{-1}f$ its Fourier transform and its inverse Fourier transform, respectively. Then $\mathcal{F}^{-1}\varphi_j\mathcal{F}f$ is an analytic function on $\mathbb{R}^n$.

Definition 7.1 *Let* $s \in \mathbb{R}$, $0 < q \le \infty$, *and let* $\{\varphi_j\}$ *be a smooth dyadic resolution of unity.*

(i) *Let* $0 < p \le \infty$. *The space* $B_{p,q}^s$ *is the collection of all* $f \in \mathcal{S}'$ *such that*

$$\|f|B_{p,q}^s\| = \Big(\sum_{j=0}^{\infty} 2^{jsq} \big\| \mathcal{F}^{-1} \varphi_j \mathcal{F}f | L_p \big\|^q \Big)^{1/q}$$

(with the usual modification if $q = \infty$*) is finite.*

(ii) *Let* $0 < p < \infty$. *The space* $F_{p,q}^s$ *is the collection of all* $f \in \mathcal{S}'$ *such that*

$$\|f|F_{p,q}^s\| = \Big\| \Big(\sum_{j=0}^{\infty} 2^{jsq} |\mathcal{F}^{-1} \varphi_j \mathcal{F}f(\cdot)|^q \Big)^{1/q} | L_p \Big\|$$

(with the usual modification if $q = \infty$*) is finite.*

Let us introduce the number

$$\sigma_p = n\left(\frac{1}{p} - 1\right)_+, \quad 0 < p \le \infty. \tag{7.2}$$

Remark 7.2   The theory of the spaces $B_{p,q}^s$ and $F_{p,q}^s$ as given above has been developed in detail in [Tri83] and [Tri92] (and continued and extended in the more recent monographs [Tri97], [Tri01], [Tri06]), but has a longer history already including many contributors; we do not want to discuss this here. Let us mention instead that these two scales $B_{p,q}^s$ and $F_{p,q}^s$ cover (fractional) Sobolev spaces, Hölder-Zygmund spaces, local Hardy spaces, and classical Besov spaces – characterised via derivatives and differences: Let $0 < p \le \infty$, $s > \sigma_p$, $0 < q \le \infty$, and $r \in \mathbb{N}$ with $r > s$. Then with $w_r(f,t)_p$ given by (2.21),

$$\|f|B_{p,q}^s\| \quad \sim \quad \|f|L_p\| + \left(\int_0^{\frac{1}{2}} \left[t^{-s} w_r(f,t)_p\right]^q \frac{dt}{t}\right)^{1/q} \tag{7.3}$$

(with the usual modification if $q = \infty$), see [BS88, Ch. 5, Def. 4.3], [DL93, Ch. 2, §10] (where the Besov spaces are defined in that way) for the Banach case, and [Tri83, Thm. 2.5.12] for what concerns the equivalence of Definition 7.1(i) and characterisation (7.3). In particular, with $p = q = \infty$, one recovers Hölder-Zygmund spaces $\mathcal{C}^s$,

$$B_{\infty,\infty}^s = \mathcal{C}^s, \quad s > 0. \tag{7.4}$$

see Definition 2.13. Let, say, $0 < s < 1$; then (in the sense of equivalent norms),

$$\|f|B_{\infty,\infty}^s\| \sim \|f|C\| + \sup_{0<t<1} \frac{w(f,t)}{t^s}. \tag{7.5}$$

Concerning F-spaces, one has, for instance, $F_{p,2}^s = H_p^s$, $s \in \mathbb{R}$, $1 < p < \infty$, the latter being the well-known (fractional) Sobolev spaces of all measurable functions $f : \mathbb{R}^n \longrightarrow \mathbb{C}$, normed by $\|f|H_p^s\| = \|I_s f|L_p\|$, where

$$I_\sigma f = \mathcal{F}^{-1}\langle\xi\rangle^\sigma \mathcal{F} f, \quad f \in \mathcal{S}', \quad \sigma \in \mathbb{R}, \tag{7.6}$$

is the lift operator; in particular, in case of classical Sobolev spaces W_p^k we have for $k \in \mathbb{N}_0, 1 < p < \infty$,

$$F_{p,2}^k = W_p^k, \quad \text{i.e.,} \quad F_{p,2}^0 = L_p. \tag{7.7}$$

Convention.   In the sequel we shall sometimes write $A_{p,q}^s$, when both scales of spaces – either $A_{p,q}^s = B_{p,q}^s$ or $A_{p,q}^s = F_{p,q}^s$ – are concerned simultaneously and the particular choice does not matter.

The lift operator I_σ, given by (7.6), maps $A_{p,q}^s$ isomorphically onto $A_{p,q}^{s-\sigma}$, $0 < p \le \infty$ (with $p < \infty$ in F-case), $0 < q \le \infty$, $s, \sigma \in \mathbb{R}$,

$$\left\| I_\sigma f | A_{p,q}^{s-\sigma} \right\| \sim \left\| f | A_{p,q}^s \right\|. \tag{7.8}$$

Another useful characterisation is given in terms of a Sobolev-type norm,

$$\left\| f | A_{p,q}^s \right\| \sim \sum_{|\alpha| \le m} \left\| D^\alpha f | A_{p,q}^{s-m} \right\| \sim \left\| f | A_{p,q}^{s-m} \right\| + \sum_{j=1}^n \left\| \frac{\partial^m f}{\partial x_j^m} \Big| A_{p,q}^{s-m} \right\|, \tag{7.9}$$

where $s \in \mathbb{R}$, $0 < p, q \le \infty$ (with $p < \infty$ in F-case), and $m \in \mathbb{N}$; see [Tri83, Thm. 2.3.8].

Remark 7.3 Note that

$$h_p = F_{p,2}^0, \quad 0 < p < \infty, \tag{7.10}$$

cf. [Tri83, Thm. 2.5.8/1]; here the local (non-homogeneous) Hardy spaces h_p, $0 < p < \infty$, are defined in the sense of Goldberg [Gol79a], [Gol79b], see also [Tri83, Sect. 2.2.2]. We shall also need the dual space of h_1, bmo $= (h_1)'$, see [Gol79b]: the local (non-homogeneous) space of functions of bounded mean oscillation, bmo, consists of all locally integrable functions $f \in L_1^{\mathrm{loc}}$ satisfying that

$$\| f | \mathrm{bmo} \| = \sup_{|Q| \le 1} \frac{1}{|Q|} \int_Q |f(x) - f_Q| \mathrm{d}x + \sup_{|Q| > 1} \frac{1}{|Q|} \int_Q |f(x)| \mathrm{d}x \tag{7.11}$$

is finite, where Q are cubes in $\mathbb{R}^n$, and f_Q is the mean value of f with respect to Q, $f_Q = \frac{1}{|Q|} \int_Q f(x) \mathrm{d}x$. This definition coincides with [Tri83, 2.2.2(viii)]; see also [BS88, Ch. 5, Def. 7.6, (7.15)].

Remark 7.4 Occasionally we shall refer in what follows to function spaces of generalised smoothness: by this we essentially mean in this context spaces of type $A_{p,q}^s$ where the main smoothness, characterised by the parameter $s \in \mathbb{R}$, is "disturbed" by some *slowly varying function* Ψ (in Karamata's sense), i.e., a positive and measurable function defined on $(0,1]$ such that

$$\lim_{t \to 0} \frac{\Psi(st)}{\Psi(t)} = 1, \quad s \in (0,1]. \tag{7.12}$$

In particular, a so-called *admissible function*, that is, a positive monotone function defined on $(0,1]$ such that $\Psi(2^{-2j}) \sim \Psi(2^{-j})$, $j \in \mathbb{N}$, is up to equivalence a slowly varying function. The standard example one should bear in mind is $\Psi_b(t) = (1 + |\log t|)^b$, $t \in (0,1]$, $b \in \mathbb{R}$. The corresponding function spaces of type $B_{p,q}^{(s,\Psi)}$ and $F_{p,q}^{(s,\Psi)}$ are introduced completely parallel to Definition 7.1 simply replacing the term 2^{js} by $2^{js}\,\Psi(2^{-j})$. These spaces were studied by Edmunds and Triebel in [ET98], [ET99a] and also considered by Moura in [Mou01], [Mou02] when Ψ is an admissible function. For further basic properties, like the independence of the spaces from the chosen dyadic resolution of unity (in the sense of equivalent norms), we refer to [FL06] in a more general setting; the extensive Russian literature can be found in the survey by Kalyabin and Lizorkin [KL87] and the appendix [Liz86]. Our intention to allude to these spaces results from connected envelope results obtained by, and partly in joint work with Caetano and Moura. This more general setting surprisingly offered deeper insight in our results; we shall explain it in more detail below.

Example 7.5 If $b \in \mathbb{R}$, then

$$\Psi_b(t) = (1 + |\log t|)^b, \quad t \in (0,1], \quad b \in \mathbb{R}, \tag{7.13}$$

is an admissible function. With this particular choice we obtain spaces $B_{p,q}^{s,b}$ consisting of those $f \in \mathcal{S}'$ for which

$$\|f \mid B_{p,q}^{s,b}\| = \Big(\sum_{j=0}^{\infty} 2^{jsq} \, (1+j)^{bq} \, \big\| \mathcal{F}^{-1} \, \varphi_j \mathcal{F} f | L_p \big\|^q \Big)^{1/q} \tag{7.14}$$

is finite (usual modification for $q = \infty$). These spaces were studied by Leopold in [Leo98], [Leo00].

It turns out that the following atomic characterisation of function spaces of type $B_{p,q}^s$ (or $F_{p,q}^s$) is sometimes preferred (compared with the above Fourier-analytical approach), e.g., when arguments for entropy numbers of embeddings between such function spaces can thus be transferred to related questions of embeddings in (well-adapted) sequence spaces; we closely follow the presentation in [Tri97, Sect. 13].

Definition 7.6 *Let* $0 < p \le \infty$, $0 < q \le \infty$, *and* $\lambda = \{\lambda_{\nu m} \in \mathbb{C} : \nu \in \mathbb{N}_0, \, m \in \mathbb{Z}^n\}$. *Then*

$$b_{pq} = \Big\{ \lambda : \|\lambda \mid b_{pq}\| = \Big(\sum_{\nu=0}^{\infty} \Big(\sum_{m \in \mathbb{Z}^n} |\lambda_{\nu m}|^p \Big)^{q/p} \Big)^{1/q} < \infty \Big\}$$

(with the usual modification if $p = \infty$ *and/or* $q = \infty$*).*

This definition is a modification of the related one in [FJ90] and coincides with [Tri97, Def. 13.5].

Let $Q_{\nu m}$, $\nu \in \mathbb{N}_0$, $m \in \mathbb{Z}^n$, denote a cube in $\mathbb{R}^n$ with sides parallel to the axes of coordinates, centred at $2^{-\nu}m$, and with side length $2^{-\nu}$. For a cube Q in $\mathbb{R}^n$ and $r > 0$ we shall mean by rQ the cube in $\mathbb{R}^n$ concentric with Q and with side length r times the side length of Q.

Definition 7.7

(i) *Let* $K \in \mathbb{N}_0$ *and* $d > 1$. *A* K *times differentiable complex-valued function* a *on* $\mathbb{R}^n$ *(continuous if* $K = 0$*) is called a* 1_K*-atom if*

$$\operatorname{supp} a \subset dQ_{0m} \quad \textit{for some} \quad m \in \mathbb{Z}^n, \tag{7.15}$$

and

$$|D^\alpha a(x)| \leq 1 \quad \textit{for} \quad |\alpha| \leq K. \tag{7.16}$$

(ii) *Let* $s \in \mathbb{R}$, $0 < p \leq \infty$, $K \in \mathbb{N}_0$, $L + 1 \in \mathbb{N}_0$, *and* $d > 1$. *A* K *times differentiable complex-valued function* a *on* $\mathbb{R}^n$ *(continuous if* $K = 0$*) is called an* $(s,p)_{K,L}$*- atom if for some* $\nu \in \mathbb{N}_0$

$$\operatorname{supp} a \subset dQ_{\nu m} \quad \textit{for some} \quad m \in \mathbb{Z}^n, \tag{7.17}$$

$$|D^\alpha a(x)| \leq 2^{-\nu(s-n/p)+|\alpha|\nu} \quad \textit{for} \quad |\alpha| \leq K, \tag{7.18}$$

and

$$\int_{\mathbb{R}^n} x^\beta a(x)\, dx = 0 \quad \textit{if} \quad |\beta| \leq L. \tag{7.19}$$

This definition coincides with [Tri97, Def. 13.3]. It is convenient to write $a_{\nu m}(x)$ instead of $a(x)$ if this atom is located at $Q_{\nu m}$ according to (7.15) and (7.17). The number d in (7.15) and (7.17) is unimportant in so far as it simply makes clear that at the level ν some controlled overlapping of the supports of $a_{\nu m}$ must be allowed. Assumption (7.19) is called a *moment condition*, where $L = -1$ means that there are no moment conditions. The atomic characterisation of function spaces of type $B^s_{p,q}$ is given by the following result [Tri97, Thm. 13.8].

Theorem 7.8 *Let* $0 < p \leq \infty$, $0 < q \leq \infty$, *and* $s \in \mathbb{R}$. *Let* $K \in \mathbb{N}_0$ *and* $L + 1 \in \mathbb{N}_0$ *with*

$$K \geq (1 + [s])_+ \quad \textit{and} \quad L \geq \max(-1, [\sigma_p - s]) \tag{7.20}$$

be fixed. Then $f \in \mathcal{S}'$ *belongs to* $B^s_{p,q}$ *if, and only if, it can be represented as*

$$f = \sum_{\nu=0}^\infty \sum_{m \in \mathbb{Z}^n} \lambda_{\nu m}\, a_{\nu m}(x), \quad \textit{convergence being in } \mathcal{S}', \tag{7.21}$$

where the $a_{\nu m}$ *are* 1_K*-atoms* $(\nu = 0)$ *or* $(s,p)_{K,L}$*-atoms* $(\nu \in \mathbb{N})$ *with*

$$\operatorname{supp} a_{\nu m} \subset dQ_{\nu m}, \quad \nu \in \mathbb{N}_0, \ m \in \mathbb{Z}^n, \ d > 1, \tag{7.22}$$

and $\lambda \in b_{pq}$. Furthermore

$$\inf \|\lambda \,|\, b_{pq}\|,$$

where the infimum is taken over all admissible representations (7.21), is an equivalent quasi-norm in $B^s_{p,q}$.

For the proof, historical comments, as well as further remarks and consequences we refer to [Tri97]; there is also a counterpart for spaces of type $F^s_{p,q}$, but this will not be needed in our context.

Remark 7.9 Dealing with F-spaces, the case $p = \infty$ is usually excluded; however they were introduced already in [Tri78b, 2.5.1] for $1 < q \le \infty$, see also [Tri83, Sect. 2.3.4]. This definition was modified and extended to $0 < q \le \infty$ by Frazier and Jawerth in [FJ90, Sect. 5]: $f \in \mathcal{S}'(\mathbb{R}^n)$ belongs to $F^s_{\infty,q}$, $s \in \mathbb{R}$, $0 < q \le \infty$, if

$$\|f|F^s_{\infty,q}\| = \sup_{Q_{\nu m}} \left(2^{\nu n} \int_{Q_{\nu m}} \left(\sum_{k=\nu}^{\infty} 2^{ksq} \left|\left(\mathcal{F}^{-1}\varphi_k \mathcal{F}f\right)(x)\right|^q \right) \mathrm{d}x \right)^{\frac{1}{q}} < \infty, \quad (7.23)$$

where the supremum is taken over all dyadic cubes $Q_{\nu m}$, $m \in \mathbb{Z}^n$, $\nu \in \mathbb{N}_0$.

Finally, in the case of a bounded domain $\Omega \subset \mathbb{R}^n$, the spaces $A^s_{p,q}(\Omega)$ are defined by restriction,

$$A^s_{p,q}(\Omega) = \{f \in D'(\Omega) : \exists\, g \in A^s_{p,q}(\mathbb{R}^n), g_{|\Omega} = f\}, \quad (7.24)$$

with $\|f|A^s_{p,q}(\Omega)\| = \inf \|g|A^s_{p,q}(\mathbb{R}^n)\|$, and the infimum is taken over all $g \in A^s_{p,q}(\mathbb{R}^n)$, $g_{|\Omega} = f$.

7.2 Embeddings

For convenience, we briefly collect some well-known facts about so-called (*non-*) *limiting* embeddings. As already explained in Part I, Sobolev's famous embedding result, Theorem 2.28, led to a large number of further embedding results in more general function spaces, say, of type $A^s_{p,q}$. Inasmuch as there is no difference we shall again omit Ω or $\mathbb{R}^n$ in the formulation below. Let $A^s_{p,q}$ stand for $B^s_{p,q}$ or $F^s_{p,q}$, respectively.

Proposition 7.10 *Let $s \in \mathbb{R}$, $0 < p \le \infty$ (with $p < \infty$ for F-spaces), and $0 < q \le \infty$.*
(i) *Then*

$$A^s_{p,q} \hookrightarrow A^s_{p,r} \quad \text{for } q \le r \le \infty. \quad (7.25)$$

(ii) *Let $0 < r \le \infty$ and $\varepsilon > 0$, then*

$$A_{p,q}^{s+\varepsilon} \hookrightarrow A_{p,r}^{s}. \tag{7.26}$$

(iii) *Assume $0 < p < \infty$, then*

$$B_{p,\min(p,q)}^{s} \hookrightarrow F_{p,q}^{s} \hookrightarrow B_{p,\max(p,q)}^{s}. \tag{7.27}$$

P r o o f : This result is well-known and can be found, for instance, in [Tri83, Prop. 2.3.2/2]. We include a short proof here for the convenience of non-specialists, to demonstrate the interplay of the different L_p- and ℓ_q-norms involved in the definition of spaces $A_{p,q}^{s}$.

In view of Definition 7.1, embedding (7.25) is a direct consequence of the monotonicity of ℓ_u-spaces, i.e., $\ell_q \hookrightarrow \ell_r$ whenever $q \le r \le \infty$. Similarly, for arbitrary $0 < r \le \infty$ and $\varepsilon > 0$ we can argue

$$\|f|B_{p,r}^{s}\| = \Big(\sum_{j=0}^{\infty} 2^{jsr} \big\| \mathcal{F}^{-1} \varphi_j \mathcal{F} f | L_p \big\|^r \Big)^{1/r}$$

$$\le \sup_{j} \, 2^{j(s+\varepsilon)} \big\| \mathcal{F}^{-1} \varphi_j \mathcal{F} f | L_p \big\| \, \Big(\sum_{j=0}^{\infty} 2^{-j\varepsilon r} \Big)^{1/r}$$

$$\le c \, \big\| f | B_{p,\infty}^{s+\varepsilon} \big\|,$$

which together with (i) finishes the proof in the B-case, with obvious modifications for $r = \infty$. The argument for the F-case works in a parallel way.

Concerning (iii), we first consider $q < p$; thus (7.27) reduces to $B_{p,q}^{s} \hookrightarrow F_{p,q}^{s}$, since the right-hand embedding already follows from $F_{p,p}^{s} = B_{p,p}^{s}$. Let $f \in B_{p,q}^{s}$, then

$$\|f|F_{p,q}^{s}\| = \Big(\int_{\mathbb{R}^n} \Big(\sum_{j=0}^{\infty} 2^{jsq} |\mathcal{F}^{-1} \varphi_j \mathcal{F} f(x)|^q \Big)^{p/q} \, \mathrm{d}x \Big)^{1/p}$$

$$= \Big(\int_{\mathbb{R}^n} \Big(\sum_{j=0}^{\infty} 2^{jsq} |\mathcal{F}^{-1} \varphi_j \mathcal{F} f(x)|^q \Big)^{u} \, \mathrm{d}x \Big)^{1/(uq)}$$

$$\le \Big(\sum_{j=0}^{\infty} \Big(\int_{\mathbb{R}^n} 2^{jsqu} |\mathcal{F}^{-1} \varphi_j \mathcal{F} f(x)|^{qu} \, \mathrm{d}x \Big)^{1/u} \Big)^{1/q}$$

$$= \Big(\sum_{j=0}^{\infty} 2^{jsp} \big\| \mathcal{F}^{-1} \varphi_j \mathcal{F} f | L_p \big\|^q \Big)^{1/q} = \|f|B_{p,q}^{s}\|,$$

where we applied, in addition, the generalised triangle inequality to $u = \frac{p}{q} > 1$, cf. [HLP52, Thm. 202]. Conversely, when $p < q$, similar arguments lead to

$$\|f|B_{p,q}^s\| = \left(\sum_{j=0}^{\infty} 2^{jsq} \left(\int_{\mathbb{R}^n} |\mathcal{F}^{-1} \varphi_j \mathcal{F} f(x)|^p \, dx \right)^{q/p} \right)^{1/q}$$

$$= \left(\sum_{j=0}^{\infty} \left(\int_{\mathbb{R}^n} 2^{jsp} |\mathcal{F}^{-1} \varphi_j \mathcal{F} f(x)|^p \, dx \right)^{v} \right)^{1/(vp)}$$

$$\leq \left(\int_{\mathbb{R}^n} \left(\sum_{j=0}^{\infty} 2^{jspv} |\mathcal{F}^{-1} \varphi_j \mathcal{F} f(x)|^{pv} \, dx \right)^{1/v} \right)^{1/p}$$

$$= \left(\int_{\mathbb{R}^n} \left(\sum_{j=0}^{\infty} 2^{jsq} |\mathcal{F}^{-1} \varphi_j \mathcal{F} f(x)|^q \right)^{p/q} \, dx \right)^{1/p} = \|f|F_{p,q}^s\|,$$

where we used $v = \frac{q}{p} > 1$. Due to (7.27) and $B_{p,p}^s = F_{p,p}^s$ this completes the proof. $\qquad\square$

Dealing with classical spaces such as L_p, $1 \leq p \leq \infty$, or C, one can complement (7.7) by the following elementary considerations. Let $f \in B_{p,1}^0$, then by the properties of a smooth dyadic resolution of unity $(\varphi_k)_k$,

$$\|f|L_p\| = \left(\int_{\mathbb{R}^n} \left| \sum_{j=0}^{\infty} \mathcal{F}^{-1} \varphi_j \mathcal{F} f(x) \right|^p dx \right)^{1/p}$$

$$\leq \sum_{j=0}^{\infty} \left(\int_{\mathbb{R}^n} |\mathcal{F}^{-1} \varphi_j \mathcal{F} f(x)|^p \, dx \right)^{1/p} = \|f|B_{p,1}^0\|,$$

with obvious modifications if $p = \infty$. Moreover, every $\mathcal{F}^{-1} \varphi_j \mathcal{F} f(x)$, $j \in \mathbb{N}_0$, is bounded and uniformly continuous on $\mathbb{R}^n$, such that the above argument implies $f \in C$, too. Conversely, for $1 \leq p \leq \infty$, we can rewrite $\mathcal{F}^{-1} \varphi_j \mathcal{F} f(x)$, $j \in \mathbb{N}$, using the properties of the (inverse) Fourier transform,

$$\mathcal{F}^{-1} \varphi_j \mathcal{F} f(x) = \int_{\mathbb{R}^n} \left(\mathcal{F}^{-1} \varphi_j \right)(y) f(x - y) \, dy, \quad x \in \mathbb{R}^n, \quad j \in \mathbb{N},$$

such that the generalised triangle inequality (for integrals), cf. [HLP52, Thm. 202], implies

$$\left\| \mathcal{F}^{-1} \varphi_j \mathcal{F} f|L_p \right\| \leq \|f|L_p\| \int_{\mathbb{R}^n} \left| \left(\mathcal{F}^{-1} \varphi_j \right)(y) \right| dy \leq c \, \|f|L_p\|$$

uniformly for all $j \in \mathbb{N}$, where we applied the special structure of (7.1), i.e., $\varphi_j(y) = \varphi_1 \left(2^{-j+1} y \right)$, $j \in \mathbb{N}$, $y \in \mathbb{R}^n$. Thus the above calculations lead to

$$B_{p,1}^0 \hookrightarrow L_p \hookrightarrow B_{p,\infty}^0, \quad 1 \leq p \leq \infty, \qquad (7.28)$$

and

$$B^0_{\infty,1} \hookrightarrow C \hookrightarrow B^0_{\infty,\infty}. \tag{7.29}$$

In view of (7.9) one obtains the following "lifted" embeddings, see [Tri83, Prop. 2.5.7, (2.5.7/10,11)].

Proposition 7.11 *Let* $m \in \mathbb{N}_0$, $1 \leq p < \infty$, *then*

$$B^m_{p,1} \hookrightarrow W^m_p \hookrightarrow B^m_{p,\infty}, \tag{7.30}$$

and

$$B^m_{\infty,1} \hookrightarrow C^m \hookrightarrow B^m_{\infty,\infty}. \tag{7.31}$$

Dealing with spaces with different metrics, we have for $0 < p_1 < p_2 < \infty$, $0 < q_1, q_2, q \leq \infty$ and $s_1 - \frac{n}{p_1} = s_2 - \frac{n}{p_2}$,

$$B^{s_1}_{p_1,q} \hookrightarrow B^{s_2}_{p_2,q} \quad \text{and} \quad F^{s_1}_{p_1,q_1} \hookrightarrow F^{s_2}_{p_2,q_2}. \tag{7.32}$$

Concerning B-spaces this is essentially due to some Plancherel-Polya-Nikolskij inequality (cf. [Tri83, (1.3.2/5), Rem. 1.4.1/4]), that gives in our case

$$\left\| \mathcal{F}^{-1} \varphi_j \mathcal{F} f | L_{p_2} \right\| \leq c \, 2^{jn \left(\frac{1}{p_1} - \frac{1}{p_2} \right)} \left\| \mathcal{F}^{-1} \varphi_j \mathcal{F} f | L_{p_1} \right\|, \quad j \in \mathbb{N}_0, \tag{7.33}$$

whereas the argument in the F-case needs some more care; we refer to [Tri83, Thm. 2.7.1].

Let us introduce the notation

$$\delta = \left(s_1 - \frac{n}{p_1} \right) - \left(s_2 - \frac{n}{p_2} \right). \tag{7.34}$$

Note that we thus have

$$A^{s_1}_{p_1,q_1} \hookrightarrow A^{s_2}_{p_2,q_2} \tag{7.35}$$

for all admitted parameters $0 < q_1, q_2 \leq \infty$, assuming that $s_1 > s_2$, $0 < p_1 \leq p_2 \leq \infty$ (with $p_2 < \infty$ in the F-case), and $\delta > 0$, whereas this is not true for $\delta = 0$ and all q-parameters in the B-case, see (7.32). We return to this point in Section 11.1. This is some reason to call $\delta = 0$ as a *limiting case* referring back to Sobolev's famous result.

In the sequel we shall need a counterpart of (7.32) concerning the case when both B- as well as F-spaces are involved (as source or target spaces, respectively). Having different smoothness parameters s_i in the spaces under consideration, then the situation (7.27) is improved as follows; we gain from a result of Sickel and Triebel in [ST95, Thm. 3.2.1].

Proposition 7.12 *Let* $0 < p_0 < p < p_1 \le \infty$, $s, s_0, s_1 \in \mathbb{R}$, *with*

$$s_0 - \frac{n}{p_0} = s - \frac{n}{p} = s_1 - \frac{n}{p_1}, \tag{7.36}$$

and $0 < q, u, v \le \infty$. *Then*

$$B^{s_0}_{p_0,u} \hookrightarrow F^s_{p,q} \hookrightarrow B^{s_1}_{p_1,v} \tag{7.37}$$

if, and only if,

$$0 < u \le p \le v \le \infty. \tag{7.38}$$

For a proof we refer to [ST95, Sect. 5.2]. Note that the "if"-part of the right-hand embedding is due to Jawerth [Jaw77], whereas the "if"-part of the left-hand embedding was proved by Franke [Fra86]. In particular,

$$B^{s_0}_{p_0,p} \hookrightarrow F^s_{p,q} \hookrightarrow B^{s_1}_{p_1,p} \tag{7.39}$$

for $0 < p_0 < p < p_1 \le \infty$, $s \in \mathbb{R}$, $s_0 - \frac{n}{p_0} = s - \frac{n}{p} = s_1 - \frac{n}{p_1}$, and $0 < q \le \infty$.

Our main goal in the next sections is to study growth envelopes or continuity envelopes of spaces of type $A^s_{p,q}$, respectively. Hence we are only interested in spaces of "functions" (i.e., regular distributions), such that $A^s_{p,q} \not\hookrightarrow L_\infty$ (in case of growth envelopes), or satisfying $A^s_{p,q} \hookrightarrow C$ and $A^s_{p,q} \not\hookrightarrow \mathrm{Lip}^1$, respectively. We discuss necessary and sufficient conditions for both cases now.

Proposition 7.13 *Let* $0 < p \le \infty$ *(with* $p < \infty$ *for F-spaces), and* $0 < q \le \infty$. *Then,*

$$F^{n/p}_{p,q} \hookrightarrow L_\infty \qquad \text{if, and only if,} \qquad 0 < p \le 1, \quad 0 < q \le \infty, \tag{7.40}$$

and

$$B^{n/p}_{p,q} \hookrightarrow L_\infty \qquad \text{if, and only if,} \qquad 0 < p \le \infty, \quad 0 < q \le 1, \tag{7.41}$$

where L_∞ *in (7.40) and (7.41) can be replaced by* C.

For a proof we refer to [ET96, 2.3.3(iii)]. Obviously, (7.40) extends Theorem 2.28(ii) due to (7.7). Moreover, by (7.26), we obtain

$$A^s_{p,q} \hookrightarrow L_\infty, \qquad s > \frac{n}{p}, \quad 0 < p, q \le \infty, \tag{7.42}$$

(with $p < \infty$ in F-case), where L_∞ can be replaced by C, too. In view of Proposition 3.4(iii) it is clear that the spaces given by (7.40), (7.41), and

(7.42), respectively, are of no further interest in our context, because the corresponding (growth) envelope functions are bounded.

In the usual $(\frac{1}{p}, s)$-diagram, where any space of the above type is characterised by its parameters s and p (neglecting q for the moment), that is $A_{p,q}^s \leftrightarrow (\frac{1}{p}, s)$, these embeddings correspond to embeddings along lines with slope n, i.e., $s - \frac{n}{p} \equiv const$. Note that this is the enriched version of Figure 8.

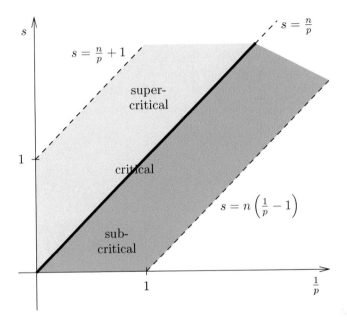

Figure 11

In view of the historical background and our above considerations, that is, the question whether a space contains essentially unbounded functions, it is reasonable to call embeddings (or spaces) of type $A_{p,q}^s$ with $s - \frac{n}{p} = 0$ *"critical"*, whereas situations with $s - \frac{n}{p} > 0$ and $s - \frac{n}{p} < 0$ are regarded as *"super-critical"* or *"sub-critical"*, respectively.

Moreover, as indicated in Figure 11, we shall merely study spaces where $\sigma_p \leq s \leq \frac{n}{p} + 1$. The lower bound is connected with the assumption $A_{p,q}^s \subset L_1^{\mathrm{loc}}$ that we have to impose, i.e., that we deal with locally integrable functions. This implies that we have to assume $s \geq \sigma_p$; a complete treatment of this problem $A_{p,q}^s \subset L_1^{\mathrm{loc}}$ can be found in [ST95, Thm. 3.3.2], where Sickel and Triebel obtained the following result:

$$F_{p,q}^s \subset L_1^{\mathrm{loc}} \iff \begin{cases} \text{either } 0 < p < 1\,,\; s \geq \sigma_p,\, 0 < q \leq \infty, \\ \text{or } 1 \leq p < \infty,\, s > \sigma_p,\, 0 < q \leq \infty, \\ \text{or } 1 \leq p < \infty,\, s = \sigma_p,\, 0 < q \leq 2\,. \end{cases} \qquad (7.43)$$

The parallel assertion for B-spaces reads as

$$B^s_{p,q} \subset L^{loc}_1 \iff \begin{cases} \text{either } 0 < p \le \infty, \ s > \sigma_p, \ 0 < q \le \infty, \\ \text{or} \quad 0 < p \le 1, \ s = \sigma_p, \ 0 < q \le 1, \\ \text{or} \quad 1 < p \le \infty, \ s = \sigma_p, \ 0 < q \le \min(p,2). \end{cases} \tag{7.44}$$

Summarising our above observations we thus concentrate on spaces

$$A^s_{p,q}, \qquad \sigma_p \le s \le \frac{n}{p}, \quad 0 < p, q \le \infty,$$

in connection with growth envelopes.

As for the study of continuity envelopes we already know by Proposition 7.13 and (7.42) that we shall assume $s \ge \frac{n}{p}$ in general. On the other hand, spaces with $s > \frac{n}{p} + 1$ are not very interesting in our context due to the following result.

Proposition 7.14 *Let* $0 < p \le \infty$ *(with $p < \infty$ for F-spaces), and* $0 < q \le \infty$. *Then*

$$F^{1+n/p}_{p,q} \hookrightarrow \text{Lip}^1 \quad \text{if, and only if,} \quad 0 < p \le 1 \quad \text{and} \quad 0 < q \le \infty, \tag{7.45}$$

and

$$B^{1+n/p}_{p,q} \hookrightarrow \text{Lip}^1 \quad \text{if, and only if,} \quad 0 < p \le \infty \quad \text{and} \quad 0 < q \le 1. \tag{7.46}$$

This is the (lifted) counterpart of (7.40) and (7.41). For a proof we refer to [Tri01, Thm. 11.4] and [EH99, Thm. 2.1]. Again, (7.26) implies

$$A^s_{p,q} \hookrightarrow \text{Lip}^1, \qquad s > \frac{n}{p} + 1, \quad 0 < p, q \le \infty, \tag{7.47}$$

(with $p < \infty$ in F-case). Hence, in view of Proposition 5.3(ii) it is clear that spaces given by (7.45), (7.46), and (7.47), respectively, are of no further interest in our context, because the corresponding envelope functions are bounded; we refer back to Section 5.1. Thus we shall rely on the notation as indicated in Figure 11, where both the *super-critical* and the *sub-critical* case are represented by the corresponding strips in the diagram.

We shall briefly dwell upon *sharp* embeddings of spaces $A^s_{p,q}$ into spaces of Lipschitz type, including the more general scale $\text{Lip}^{(1,-\alpha)}_{p,q}$ introduced in Section 2.3. Since we have already reserved the expression "limiting" (in connection with embeddings) for situations described by (7.34), we shall adopt the saying *"sharp embedding"* now when – at least for one parameter – there cannot be chosen any "better" (smaller or larger, respectively) value such that the embedding still holds. For instance, returning to the famous result of Brézis and Wainger [BW80], see (1.13) and rewritten now as

$$H_p^{1+n/p} \hookrightarrow \mathrm{Lip}^{(1,-1/p')}, \tag{7.48}$$

see also (1.14), (2.60), one can ask whether the embedding (7.48) is *sharp* in the sense that

$$H_p^{1+n/p} \not\hookrightarrow \mathrm{Lip}^{(1,-\alpha)}$$

if $\alpha < \frac{1}{p'}$ (by the monotonicity of spaces $\mathrm{Lip}^{(1,-\alpha)}$ in α one clearly looks for the smallest value of α).

We formulate our result in terms of spaces $\mathrm{Lip}_{p,q}^{(1,-\alpha)}$ introduced in Definition 2.16.

Proposition 7.15 *Let* $0 < q, v \leq \infty$, $\alpha > \frac{1}{v}$ *(with* $\alpha \geq 0$ *if* $v = \infty$*).*

(i) *Let* $0 < p \leq \infty$. *Then*

$$B_{p,q}^{1+n/p} \hookrightarrow \mathrm{Lip}_{\infty,v}^{(1,-\alpha)} \qquad \text{if, and only if,} \qquad \alpha \geq \frac{1}{v} + \frac{1}{q'}. \tag{7.49}$$

In particular, for $v = \infty$,

$$B_{p,q}^{1+n/p} \hookrightarrow \mathrm{Lip}_{\infty,\infty}^{(1,-\alpha)} = \mathrm{Lip}^{(1,-\alpha)} \quad \text{if, and only if,} \quad \alpha \geq \frac{1}{q'}. \tag{7.50}$$

(ii) *Let* $0 < p < \infty$. *Then*

$$F_{p,q}^{1+n/p} \hookrightarrow \mathrm{Lip}_{\infty,v}^{(1,-\alpha)} \qquad \text{if, and only if,} \qquad \alpha \geq \frac{1}{v} + \frac{1}{p'}. \tag{7.51}$$

In particular, for $v = \infty$,

$$F_{p,q}^{1+n/p} \hookrightarrow \mathrm{Lip}_{\infty,\infty}^{(1,-\alpha)} = \mathrm{Lip}^{(1,-\alpha)} \quad \text{if, and only if,} \quad \alpha \geq \frac{1}{p'}. \tag{7.52}$$

We refer to [EH99, Thm. 2.1] for the case $v = \infty$ and to [Har02, Prop. 3.3.5] otherwise. Clearly, (7.50) and (7.52) coincide with Proposition 7.14 when $\alpha = 0$. In particular, (7.52) implies that for $1 < p < \infty$ and $0 < q \leq \infty$ there is some $c > 0$ such that for all $x, y \in \mathbb{R}^n$, $0 < |x - y| < \frac{1}{2}$, and all $f \in F_{p,q}^{1+n/p}$,

$$|f(x) - f(y)| \leq c\,|x - y|\,\left|\log|x - y|\right|^{1/p'}\left\|f|F_{p,q}^{1+n/p}\right\|, \tag{7.53}$$

where the exponent $\frac{1}{p'}$ is sharp. Similarly, for $0 < p \leq \infty$ and $1 < q \leq \infty$ there is some $c > 0$ such that for all $x, y \in \mathbb{R}^n$, $0 < |x - y| < \frac{1}{2}$, and all $f \in B_{p,q}^{1+n/p}$,

$$|f(x) - f(y)| \leq c\, |x - y| \left| \log |x - y| \right|^{1/q'} \left\| f | B_{p,q}^{1+n/p} \right\|, \qquad (7.54)$$

where the exponent $\frac{1}{q'}$ is sharp, see (7.50). Using $F_{p,2}^s = H_p^s$, $s \in \mathbb{R}$, $1 < p < \infty$, (7.51) generalises (1.13). For other works on sharpness of related embeddings see [EGO97], [EGO00] and [EK95], see also our discussion in [EH99, Rem. 2.5].

Example 7.16 Let $p = q = \infty$, such that $B_{\infty,\infty}^1 = \mathcal{C}^1$ in view (7.4). Then, according to (7.54), there is some $c > 0$ such that for all $f \in \mathcal{C}^1$,

$$|f(x) - f(0)| \leq c\, |x| \left| \log |x| \right| \|f|\mathcal{C}^1\|, \qquad (7.55)$$

for all x, $0 < |x| < \frac{1}{2}$. The exponent 1 of $\left| \log |x| \right|$ in (7.55) is sharp. This was first observed by Zygmund [Zyg45].

Remark 7.17 The sharpness assertion essentially relies on results on extremal functions as presented in [EH99, Prop. 2.2]: Let $1 < p < \infty$ and $\sigma > \frac{1}{p}$. There is a function $g_{p\sigma}$ with

$$g_{p\sigma} \in B_{p,p}^{1+n/p}, \qquad g_{p\sigma}(0) = 0,$$

$$|g_{p\sigma}(x)| \geq c\, |x| \left| \log |x| \right|^{1/p'} \left(\log \left| \log \varepsilon |x| \right| \right)^{-\sigma}$$

for some $c > 0$, small $\varepsilon > 0$ and $x = (x_1, 0, \ldots, 0)$, $0 < x_1 < \delta$, $\delta > 0$ small. This is some "lifted" version of an example given by Triebel in [Tri93, Thms. 3.1.2, 4.2.2]; see also [ET96, Thm. 2.7.1].

We want to locate the Lipschitz spaces $\mathrm{Lip}_{p,q}^{(1,-\alpha)}$ within the scale of Besov spaces, refined by the logarithmic spaces $B_{p,q}^{s,b}$ introduced in Example 7.5, see (7.14). We start with the classical setting $b = 0$.

Proposition 7.18 Let $1 \leq p < \infty$, $0 < q, v \leq \infty$, $\alpha > \frac{1}{v}$ (with $\alpha \geq 0$ if $v = \infty$). Then

$$B_{p,q}^1 \hookrightarrow \mathrm{Lip}_{p,v}^{(1,-\alpha)} \qquad if \qquad \begin{cases} \alpha \geq \frac{1}{q'}, & v = \infty, \\ \alpha > \frac{1}{v} + \frac{1}{q'}, & v < \infty. \end{cases} \qquad (7.56)$$

This result is proved and discussed in [Har00b, Prop. 11] for all p, $1 \leq p \leq \infty$, but the case $p = \infty$ is now replaced by the better result Proposition 7.15. Furthermore, for $p < \infty$ and $v = \infty$ (7.56) is covered by [EH00, Prop. 4.2(ii)] already. Comparing (7.56) and (7.49) the question naturally arises whether $B_{p,q}^1 \hookrightarrow \mathrm{Lip}_{p,v}^{(1,-\alpha)}$ remains true for $\alpha = \frac{1}{v} + \frac{1}{q'}$ and $v < \infty$, $p < \infty$. This is not so clear at the moment. However, when $p = \infty$ [Har00b, Cor. 20] implies that there cannot be an embedding like (7.56) for $\alpha < \frac{1}{v} + \frac{1}{q'}$. Otherwise, for $1 \leq p < \infty$, there is an improved version of (7.56) by Neves in [Nev01a, Prop. 5.2] based upon Timan's inequality [DL93, Ch. 2, Thm. 8.4] instead of Marchaud's (2.22).

As already mentioned, we are interested in extensions of (7.4) and

$$B_{\infty,1}^1 \hookrightarrow \mathrm{Lip}^1 \hookrightarrow B_{\infty,\infty}^1,$$

see [Tri83, (2.5.7/2), (2.5.7/11)], to spaces of type $\mathrm{Lip}^{(1,-\alpha)}$, $\mathcal{C}^{(1,-\alpha)}$. In [EH99, Props. 4.2, 4.4] we proved such a result.

Proposition 7.19 *Let $\alpha \geq 0$. Then*

$$B_{\infty,1}^{1,-\alpha} \hookrightarrow \mathrm{Lip}^{(1,-\alpha)} \hookrightarrow \mathcal{C}^{(1,-\alpha)} = B_{\infty,\infty}^{1,-\alpha}. \tag{7.57}$$

Moreover,

$$B_{\infty,q}^{1,-\alpha} \hookrightarrow \mathrm{Lip}^{(1,-\alpha)} \qquad \text{if, and only if,} \qquad 0 < q \leq 1.$$

Our so far final embedding results related to spaces $B_{p,q}^{s,b}$ and $\mathrm{Lip}_{p,q}^{(1,-\alpha)}$ is the following, collecting [Har00b, Prop. 23, Cors. 25, 26]. We shall use it later on to derive envelope assertions for spaces $\mathrm{Lip}_{p,q}^{(1,-\alpha)}$.

Corollary 7.20 *Let $1 \leq p \leq \infty$, $0 < q \leq \infty$, $\alpha > \frac{1}{q}$.*
(i) *Then*

$$B_{p,1}^{1,-\beta} \hookrightarrow \mathrm{Lip}_{p,q}^{(1,-\alpha)} \qquad \text{if} \quad \begin{cases} \beta < \alpha - \frac{1}{q}, & 0 < q < \infty, \\ \beta \leq \alpha, & q = \infty. \end{cases} \tag{7.58}$$

Moreover,

$$B_{p,\min(q,1)}^{1,-\alpha+\frac{1}{q}} \hookrightarrow \mathrm{Lip}_{p,q}^{(1,-\alpha)}. \tag{7.59}$$

(ii) *Let $1 \leq q \leq \infty$, $\alpha > 1$. Then*

$$B_{p,q}^{1,-\alpha+1} \hookrightarrow \mathrm{Lip}_{p,q}^{(1,-\alpha)}. \tag{7.60}$$

(iii) *Let* $0 < v \leq \infty$, $\beta > \frac{1}{v}$. *Then*

$$
\mathrm{Lip}_{p,q}^{(1,-\alpha)} \hookrightarrow B_{p,v}^{1,-\beta} \quad if \quad
\begin{cases}
\beta - \frac{1}{v} \geq \alpha - \frac{1}{q}, & v \geq q, \\[2mm]
\beta - \frac{1}{v} > \alpha - \frac{1}{q}, & v < q.
\end{cases}
\tag{7.61}
$$

Concerning the question *"where"* the Lipschitz spaces $\mathrm{Lip}_{p,q}^{(1,-\alpha)}$ can be found within the scale of Besov spaces $B_{p,q}^{s,b}$, we arrived in [Har00b, Sect. 4] at the following diagram, where $q^* := \min(q,1)$,

$$
\begin{array}{ccccccccc}
& & B_{p,q}^{1,-\alpha+\frac{1}{q^*}} & & & & B_{p,q}^{1,-\alpha} & & \\
& \nearrow & & \searrow & & \nearrow & & \searrow & \\
B_{p,q^*}^{1,-\alpha+\frac{1}{q^*}} & & & & \mathrm{Lip}_{p,q}^{(1,-\alpha)} & & & & B_{p,\infty}^{1,-\alpha}. \\
& \searrow & & \nearrow & & \searrow & & \nearrow & \\
& & B_{p,q^*}^{1,-\alpha+\frac{1}{q}} & & & & B_{p,\infty}^{1,-\alpha+\frac{1}{q}} & &
\end{array}
$$

$$
\underbrace{\phantom{B_{p,q^*}^{1,-\alpha+\frac{1}{q^*}}\quad\quad\quad B_{p,q^*}^{1,-\alpha+\frac{1}{q}}}}_{\displaystyle \equiv B_{p,q}^{1,-\alpha+\frac{1}{q}},\ 0 < q \leq 1}
\qquad
\underbrace{\phantom{B_{p,q}^{1,-\alpha}\quad\quad\quad B_{p,\infty}^{1,-\alpha+\frac{1}{q}}}}_{\displaystyle \equiv B_{p,\infty}^{1,-\alpha},\ q = \infty}
$$

We have the same diagram with $\mathrm{Lip}_{p,q}^{(1,-\alpha)}$ replaced by $B_{p,q}^{1,-\alpha+\frac{1}{q}}$. These spaces, however, are not comparable (in the above sense) when $1 < q < \infty$. There are a lot of further related approaches to spaces of Lipschitz type; we refer to [Har00b] for more details.

Remark 7.21 As mentioned in Remark 7.4 above, we shall occasionally refer to function spaces of generalised smoothness $A_{p,q}^{(s,\Psi)}$ in the sequel, where Ψ is a *slowly varying function* or an *admissible function*. Note that many of the above embeddings have their immediate counterpart in this context; in particular, this covers embeddings (7.25), (7.26), (7.27), (7.32), (7.39), always assuming that the same function Ψ is involved both in source and target space; cf. [Mou02, Prop. 1.1.13], [BM03], [HM04]. In addition, one has for all admitted parameters and all $\varepsilon > 0$,

$$
A_{p,q}^{s+\varepsilon} \hookrightarrow A_{p,q}^{(s,\Psi)} \hookrightarrow A_{p,q}^{s-\varepsilon}.
\tag{7.62}
$$

Concerning Propositions 7.13, 7.14 there are the following counterparts: Let $0 < p,q \leq \infty$ (with $p < \infty$ for F-spaces), and Ψ as above, then

$$
B_{p,q}^{(n/p,\Psi)} \hookrightarrow C \quad \text{if, and only if,} \quad \left(\Psi\left(2^{-j}\right)^{-1} \right)_{j \in \mathbb{N}} \in \ell_{q'}
\tag{7.63}
$$

and

$$
F_{p,q}^{(n/p,\Psi)} \hookrightarrow C \quad \text{if, and only if,} \quad \left(\Psi\left(2^{-j}\right)^{-1} \right)_{j \in \mathbb{N}} \in \ell_{p'},
\tag{7.64}
$$

where C may be replaced by L_∞. We refer to [CM04a, Prop. 3.11] for a proof; a forerunner related to the case $1 < p, q < \infty$ was obtained in [Kal82]. As for the embeddings in Lipschitz spaces, we obtained in [CH05, Prop. 2.2]

$$B_{p,q}^{(1+n/p,\Psi)} \hookrightarrow \mathrm{Lip}^1 \quad \text{if, and only if,} \quad \left(\Psi\left(2^{-j}\right)^{-1}\right)_{j\in\mathbb{N}} \in \ell_{q'}, \qquad (7.65)$$

and

$$F_{p,q}^{(1+n/p,\Psi)} \hookrightarrow \mathrm{Lip}^1 \quad \text{if, and only if,} \quad \left(\Psi\left(2^{-j}\right)^{-1}\right)_{j\in\mathbb{N}} \in \ell_{p'}. \qquad (7.66)$$

Due to (7.43), (7.44), and (7.62), the assumption

$$s > \sigma_p = n\left(\frac{1}{p} - 1\right)_+$$

already implies $A_{p,q}^{(s,\Psi)} \subset L_1^{\mathrm{loc}}$.

Example 7.22 We return to our Example 7.5 concerning spaces $B_{p,q}^{s,b}$. In view of Propositions 7.13, 7.14, (7.62), (7.63), and (7.65) we thus have

$$B_{p,q}^{s,b} \hookrightarrow L_\infty$$

whenever $s > \frac{n}{p}$, and $b \in \mathbb{R}$, $0 < q \le \infty$ arbitrary, whereas in the limiting case,

$$B_{p,q}^{n/p,b} \hookrightarrow L_\infty \quad \text{if, and only if,} \quad \left\{ \begin{array}{l} b > \frac{1}{q'} \,, 1 < q \le \infty \\ b \ge 0 \;\;, 0 < q \le 1 \end{array} \right\}.$$

Otherwise, assuming $s > \frac{n}{p}$, we obtain

$$B_{p,q}^{s,b} \hookrightarrow \mathrm{Lip}^1$$

if $s > \frac{n}{p} + 1$, and $b \in \mathbb{R}$, $0 < q \le \infty$ arbitrary, or, in the limiting case,

$$B_{p,q}^{1+n/p,b} \hookrightarrow \mathrm{Lip}^1 \quad \text{if, and only if,} \quad \left\{ \begin{array}{l} b > \frac{1}{q'} \,, 1 < q \le \infty \\ b \ge 0 \;\;, 0 < q \le 1 \end{array} \right\}. \qquad (7.67)$$

Chapter 8

Growth envelopes $\mathfrak{E}_G$ of function spaces $A_{p,q}^s$

We study growth envelopes, introduced in Chapters 3 and 4, in the context of function spaces of type $A_{p,q}^s$, where $\sigma_p \leq s \leq \frac{n}{p}$.

8.1 Growth envelopes in the sub-critical case

We first deal with spaces of type $A_{p,q}^s$ where $0 < p < \infty$, $0 < q \leq \infty$, and $\sigma_p \leq s < \frac{n}{p}$ – called *sub-critical* according to our notation in Figure 11 (and the explanations given there). The *borderline case* $s = \sigma_p$ needs some additional care; this refers to the thick lines in Figure 12. This situation is studied separately, but postponed to the end of this section.

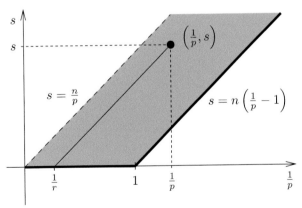

Figure 12

First we consider the *"sub-critical strip"* where $\sigma_p < s < \frac{n}{p}$, $0 < p < \infty$ and $0 < q \leq \infty$. Let $1 < r < \infty$, then all spaces on the line with slope n

and "foot-point" $\frac{1}{r}$ (see Figure 12) belong to this *sub-critical* area. Moreover, as all spaces of type $A^s_{p,q}$ (with such parameters) can be embedded in, say, suitable Lebesgue spaces L_u, it makes sense to study their growth envelopes.

Theorem 8.1 *Let* $0 < q \leq \infty$, $0 < p < \infty$, *and* $\sigma_p < s < \frac{n}{p}$. *Then*

$$\mathfrak{E}_G\left(F^s_{p,q}\right) = \left(t^{-\frac{1}{p}+\frac{s}{n}}, p\right) \tag{8.1}$$

and

$$\mathfrak{E}_G\left(B^s_{p,q}\right) = \left(t^{-\frac{1}{p}+\frac{s}{n}}, q\right). \tag{8.2}$$

P r o o f :   This result is known, see [Tri01, Thm. 15.2]; we give here a proof for later reference when we deal with borderline cases and indicate necessary modifications. Let $1 < r < \infty$ be given by $\frac{n}{p} - s = \frac{n}{r}$, as indicated in Figure 12.

Thus the above assertions read as $\mathfrak{E}_G^{A^s_{p,q}}(t) \sim t^{-\frac{1}{r}}$, with $u_G^{B^s_{p,q}} = q$, $u_G^{F^s_{p,q}} = p$.

Step 1.   We first show $\mathfrak{E}_G^{A^s_{p,q}}(t) \sim t^{-\frac{1}{r}}$ under the above assumptions. Note that by (7.7) and (7.32) we have

$$F^s_{p,q} \hookrightarrow F^0_{r,2} = L_r, \quad 1 < r < \infty. \tag{8.3}$$

Now Proposition 3.4(iv) and Theorem 4.7(i) immediately imply $\mathfrak{E}_G^{F^s_{p,q}}(t) \leq c\, t^{-\frac{1}{r}}$. Concerning the corresponding estimate for B-spaces, note that by a real interpolation argument

$$\left(F^s_{p_0,q}, F^s_{p_1,q}\right)_{\theta,p} = F^s_{p,q} \qquad \text{and} \qquad \left(L_{r_0}, L_{r_1}\right)_{\theta,p} = L_{r,p}, \tag{8.4}$$

where $0 < \theta < 1$, $1 < r_0 < r_1 < \infty$, $0 < p_0 < p_1 < \infty$, $s > 0$, and $0 < q \leq \infty$. Here

$$\frac{1}{r} = \frac{1-\theta}{r_0} + \frac{\theta}{r_1} \qquad \text{and} \qquad \frac{1}{p} = \frac{1-\theta}{p_0} + \frac{\theta}{p_1}. \tag{8.5}$$

The F-interpolation in (8.4) restricted to $1 < p_i < \infty$ and $1 < q < \infty$ coincides with [Tri78a, 2.4.2/(6)], whereas the general case is due to Frazier and Jawerth, [FJ90, Cor. 6.7 and §12]; the L-part in (8.4) is the very classical interpolation result for Lebesgue spaces, see (3.55) with $r_0 = q_0$, $r_1 = q_1$, and cf. [BL76, Thm. 5.3.1] and [Tri78a, Thm. 1.18.6/2]. Thus (8.3) together with (8.4) results in

$$F^s_{p,q} \hookrightarrow L_{r,p}. \tag{8.6}$$

The B-case now follows from another real interpolation (with $B^s_{p,p} = F^s_{p,p}$) or (8.6) with (7.39),

$$B^s_{p,q} \hookrightarrow L_{r,q}; \tag{8.7}$$

for the case p, $q \geq 1$ see also [BS88, Ch. 5, Cor. 4.20]. This finally leads to $\mathcal{E}^{B^s_{p,q}}_G(t) \leq c\, t^{-\frac{1}{r}}$, using Proposition 3.4(iv) and Theorem 4.7(i) again.

<u>Step 2.</u>   It remains to show the converse inequalities. We use the example given in [Tri99, 3.2] and return to the construction introduced in Part I in (3.35). Let $\psi(x)$ be the compactly supported C^∞-function in $\mathbb{R}^n$ given by (3.36); then for $j \in \mathbb{N}$, the functions

$$f_j(x) := 2^{j\frac{n}{r}}\, \psi\left(2^j x\right), \quad x \in \mathbb{R}^n, \tag{8.8}$$

are atoms in $B^s_{p,q}$ in the sub-critical case: one has (up to a normalising factor which can be neglected) supp $f_j \subset \{x \in \mathbb{R}^n : |x| \leq 2^{-j}\}$, and

$$|D^\gamma f_j(x)| \leq 2^{-j(s-\frac{n}{p})+j|\gamma|} = 2^{j\frac{n}{r}+j|\gamma|}, \quad |\gamma| \leq K,$$

where $K \in \mathbb{N}$ can be chosen arbitrarily large. Moreover, dealing with the sub-critical B-setting one does not need any moment conditions, as

$$\sigma_p - s = n\left(\frac{1}{p} - 1\right)_+ - s = n\left(\frac{1}{p} - 1\right)_+ - \frac{n}{p} + \frac{n}{r} < 0 \tag{8.9}$$

for any p, $0 < p < \infty$. In particular, the functions f_j, $j \in \mathbb{N}$, belong to $B^s_{p,q}$ with $\|f_j|B^s_{p,q}\| \sim 1$ (independent of $j \in \mathbb{N}$). This can be seen by Theorem 7.8. On the other hand one calculates

$$f_j^*\left(2^{-jn}\right) \sim 2^{j\frac{n}{r}}, \quad j \in \mathbb{N},$$

implying

$$\mathcal{E}^{B^s_{p,q}}_G\left(2^{-jn}\right) \geq f_j^*\left(2^{-jn}\right) \sim 2^{j\frac{n}{r}}, \quad j \in \mathbb{N}.$$

This yields the desired B-result,

$$\mathcal{E}^{B^s_{p,q}}_G(t) \geq c\, t^{-\frac{1}{r}}, \quad 0 < t < 1. \tag{8.10}$$

Moreover, choose for given s and p parameters $\sigma > s$ and $0 < v < p$ such that $\sigma - \frac{n}{v} = s - \frac{n}{p} = -\frac{n}{r}$. By (7.39), $B^\sigma_{v,p} \hookrightarrow F^s_{p,q}$; hence (8.10) and Proposition 3.4(iv) complete the lower estimate in the F-setting,

$$\mathcal{E}^{F^s_{p,q}}_G(t) \geq c\, t^{-\frac{1}{r}}, \quad 0 < t < 1.$$

<u>Step 3.</u>   Finally we have to verify that in the F-case $v = p$ is the smallest possible exponent satisfying

$$\left(\int_0^\varepsilon \left[t^{\frac{1}{r}} f^*(t)\right]^v \frac{dt}{t}\right)^{1/v} \leq c\, \|f|F^s_{p,q}\|$$

for all $f \in F_{p,q}^s$, whereas in the B-case $v = q$ is optimal. Note that we have $\mathcal{E}_\mathsf{G}^{F_{p,q}^s}(t) \sim \mathcal{E}_\mathsf{G}^{B_{v,p}^\sigma}(t)$ whenever $\sigma - \frac{n}{v} = s - \frac{n}{p} = -\frac{n}{r}$ so that by Proposition 4.5 together with (7.39) it is sufficient to deal with the B-case only. Clearly, $\mu_\mathsf{G}(dt) \sim \frac{dt}{t}$, and another application of Proposition 4.5 together with (8.7) and Theorem 4.7(i) immediately provides $u_\mathsf{G}^{B_{p,q}^s} \leq q$. The sharpness can be either taken as a consequence of [Tri99, Cor. 2.5], see also [Tri01, Thm. 15.2], or derived in a way parallel to the argument presented in the proof of Proposition 4.10: we consider a refined construction of the above "extremal" functions f_j. Let $\{b_j\}_{j\in\mathbb{N}}$ be a sequence of non-negative numbers, and let $x^0 \in \mathbb{R}^n$, $|x^0| > 4$, such that $\operatorname{supp} \psi \left(2^j \cdot -x^0 \right) \cap \operatorname{supp} \psi \left(2^r \cdot -x^0 \right) = \emptyset$ for $j \neq r$, $j, r \in \mathbb{N}_0$, and ψ given by (3.36). Then

$$f_b(x) := \sum_{j=1}^\infty 2^{j\frac{n}{r}} b_j \, \psi \left(2^j x - x^0 \right), \quad x \in \mathbb{R}^n, \tag{8.11}$$

can be seen as atomic decomposition of f_b; hence Theorem 7.8 implies

$$\left\| f_b | B_{p,q}^s \right\| \leq c \, \| b | \ell_q \|. \tag{8.12}$$

Recall $f_b^* \left(c \, 2^{-jn} \right) \geq c' \, b_j \, 2^{j\frac{n}{r}}$, $j \in \mathbb{N}_0$. For convenience we may assume $b_1 = \cdots = b_{J-1} = 0$, where J is suitably chosen such that $2^{-J} \sim \varepsilon$ given by (4.14). Then by monotonicity arguments,

$$\left(\int_0^\varepsilon \left[t^{\frac{1}{r}} f_b^*(t) \right]^v \frac{dt}{t} \right)^{\frac{1}{v}} \geq c_1 \left(\sum_{j=J}^\infty \left[2^{-j\frac{n}{r}} f_b^* \left(c2^{-jn} \right) \right]^v \right)^{\frac{1}{v}} \geq c_2 \left(\sum_{j=J}^\infty b_j^v \right)^{\frac{1}{v}},$$

i.e., we arrive at $\| b | \ell_q \| \geq c \, \| b | \ell_v \|$ for arbitrary sequences of non-negative numbers. This obviously requires $v \geq q$. $\qquad\square$

Remark 8.2 Observe that (8.1) implies Proposition 4.10 when $1 < p < \infty$ in view of (7.7). Moreover, (8.1) together with Theorem 4.7(i) lead to

$$\mathfrak{E}_\mathsf{G}\left(L_{r,p} \right) = \left(t^{-\frac{1}{r}}, \, p \right) = \mathfrak{E}_\mathsf{G}\left(F_{p,q}^s \right), \tag{8.13}$$

where $0 < q \leq \infty$, $s > 0$, $1 < r < \infty$ and $0 < p < \infty$ with $s - \frac{n}{p} = -\frac{n}{r}$; that is, we have by (8.6) the embedding $F_{p,q}^s \hookrightarrow L_{r,p}$ only, whereas the corresponding envelopes even coincide. This can be interpreted as $L_{r,p}$ being indeed the best possible space within the Lorentz scale in which $F_{p,q}^s$ can be embedded continuously. On the other hand this is to be understood in the sense that $L_{r,p}$ is "as good as" $F_{p,q}^s$ – as far as only the growth of the unbounded functions belonging to the spaces under consideration is concerned; (additional) smoothness features (making a big difference between the spaces

$L_{r,p}$ and $F_{p,q}^s$, for instance) are obviously *"ignored"* by the growth envelope. This is not really astonishing in view of its construction, but worth noticing. The parallel assertion for the B-case, i.e., (8.2) together with Theorem 4.7(i) provide

$$\mathfrak{E}_G\left(L_{r,q}\right) = \left(t^{-\frac{1}{r}}, \, q\right) = \mathfrak{E}_G\left(B_{p,q}^s\right), \tag{8.14}$$

the parameters being as above. Again we note by (8.7) that $B_{p,q}^s$ can be embedded in $L_{r,q}$, whereas their envelopes even coincide. We return to this phenomenon in Section 11.2.

The embedding result (8.7) can (in the Banach space situation) also be found in [Gol87a] and [Kol98]. Moreover, Gol'dman's result [Gol87b, Thm. 2.1, Cor. 5.1] can be interpreted as the fact that $L_{r,q}$ is the best possible space within the Lorentz scale in which $B_{p,q}^s$ can be embedded continuously – coinciding with our above interpretation of (8.13), see also [Her68, Thm. 8.5].

Remark 8.3 Forerunners of this result – formulated in a different context – are presented in [Tri99]. This is extended and generalised in [Tri01, Sect. 15]. There one also finds a lot of remarks and references on the long history of related studies; thus we shall only mention some of the most important names and papers briefly: essential contributions were achieved by Peetre [Pee66], Strichartz [Str67], Herz [Her68], as well as in the Russian school by Brudnyi [Bru72], Gol'dman [Gol87c], [Gol87b], Lizorkin [Liz86], Kalyabin, Lizorkin [KL87], Netrusov [Net87], [Net89], see also the book by Ziemer [Zie89]. More recent treatments are, for instance, [CP98] by Cwikel, Pustylnik, [EKP00] by Edmunds, Kerman, Pick and the surveys [Kol98] by Kolyada, [Tar98] by Tartar. There are far more investigations connected in some sense with limiting embeddings; we refer to the survey papers for detailed information.

Remark 8.4 Recently Caetano and Moura obtained parallel results in the sub-critical case when studying spaces of generalised smoothness of type $B_{p,q}^{(s,\Psi)}$, $F_{p,q}^{(s,\Psi)}$, see Remark 7.4. According to Remark 7.21 we assume $0 < p, q \le \infty$, $\sigma_p < s < \frac{n}{p}$. The corresponding sub-critical result in [CM04b, Thm. 4.4] then reads as

$$\mathfrak{E}_G\left(B_{p,q}^{(s,\Psi)}\right) = \left(t^{-\frac{1}{r}} \, \Psi(t)^{-1}, \, q\right) \tag{8.15}$$

and

$$\mathfrak{E}_G\left(F_{p,q}^{(s,\Psi)}\right) = \left(t^{-\frac{1}{r}} \, \Psi(t)^{-1}, \, p\right). \tag{8.16}$$

Further extensions were obtained by Bricchi and Moura in [BM03] concerning spaces $A_{p,q}^\sigma(\mathbb{R}^n)$, where the usual (scalar) regularity index $\sigma \in \mathbb{R}$ is replaced by a sequence $\boldsymbol{\sigma} = \{\sigma_j\}_{j \in \mathbb{N}_0}$. Releasing also the subordinate (dyadic) partition of unity one obtains spaces $A_{p,q}^{\sigma,N}$ of generalised smoothness; their growth envelopes were studied in [CF06], [CL06]. There is some recent work on anisotropic spaces [MNP06] that leads to the same, i.e., isotropic results.

Finally, growth envelopes of another possible modification of spaces, namely Lorentz-Karamata-Bessel potential spaces $H^s L_{p,q;\Psi}(\mathbb{R}^n)$ with a slowly varying function Ψ are investigated in [GNO04], see also [GO06].

Example 8.5 We return to Example 7.5, $s \in \mathbb{R}$, $b \in \mathbb{R}$, $0 < p < \infty$, $0 < q \leq \infty$. In view of Example 7.22 we assume $\sigma_p < s < \frac{n}{p}$, referring to the sub-critical case. Then (8.15) implies

$$\mathfrak{E}_G\left(B_{p,q}^{s,b}\right) = \left(t^{-\frac{1}{r}} |\log t|^{-b}, \, q\right) = \mathfrak{E}_G\left(L_{r,q}(\log L)_a\right) \qquad (8.17)$$

where $1 < r < \infty$ is given by $s - \frac{n}{p} = -\frac{n}{r}$; see Theorem 4.7(i). This is the counterpart of (8.14), adapted to this more general setting.

Corollary 8.6 *Let $1 \leq p < n$, $0 < q \leq \infty$, $\alpha > \frac{1}{q}$, then*

$$\mathcal{E}_G^{\mathrm{Lip}_{p,q}^{(1,-\alpha)}}(t) \sim t^{-\frac{1}{p}+\frac{1}{n}} |\log t|^{\alpha-\frac{1}{q}}, \quad 0 < t < \frac{1}{2}. \qquad (8.18)$$

P r o o f :   Note that by (7.59) and (7.61) we have

$$B_{p,\min(q,1)}^{1,-\alpha+\frac{1}{q}} \hookrightarrow \mathrm{Lip}_{p,q}^{(1,-\alpha)} \hookrightarrow B_{p,\infty}^{1,-\alpha+\frac{1}{q}}.$$

Now the result follows by Proposition 3.4(iv) and (8.17).   □

We conclude this section with a short digression on weighted spaces of type $A_{p,q}^s(w_\alpha)$. Recall our notation (3.48),

$$w_\alpha(x) = \langle x \rangle^\alpha, \quad \alpha \in \mathbb{R}, \quad x \in \mathbb{R}^n,$$

and Remark 3.29. In continuation of the spaces $L_p(\mathbb{R}^n, w) = L_p(w)$ introduced in (3.47) one can define spaces of type $B_{p,q}^s(\mathbb{R}^n, w)$ and $F_{p,q}^s(\mathbb{R}^n, w)$ completely parallel to Definition 7.1 simply replacing $L_p(\mathbb{R}^n)$ by $L_p(\mathbb{R}^n, w)$, where w is admissible in the sense of Section 3.4. In [HT94a] we have proved that $f \in A_{p,q}^s(\mathbb{R}^n, w)$ if, and only if, $wf \in A_{p,q}^s(\mathbb{R}^n)$, more precisely, that for such admissible weights w,

$$\left\| f \mid A_{p,q}^s(\mathbb{R}^n, w) \right\| \sim \left\| wf \mid A_{p,q}^s(\mathbb{R}^n) \right\| \qquad (8.19)$$

are equivalent norms. Moreover, we proved in [HT94a, Thm. 2.3] that for $A = F$,

$$A_{p_1,q_1}^{s_1}(w_1) \hookrightarrow A_{p_2,q_2}^{s_2}(w_2) \qquad (8.20)$$

if, and only if,

$$s_1 - \frac{n}{p_1} \geq s_2 - \frac{n}{p_2} \quad \text{and} \quad \frac{w_2(x)}{w_1(x)} \leq c < \infty \qquad (8.21)$$

for some $c > 0$ and all $x \in \mathbb{R}^n$. Here we assume $s_1 > s_2$, $0 < p_1 \leq p_2 < \infty$, $0 < q_1, q_2 \leq \infty$, and w_1, w_2 are admissible weight functions. For $A = B$ we got the same assertion, but in the limiting case $s_1 - \frac{n}{p_1} = s_2 - \frac{n}{p_2}$ we have to observe $q_1 \leq q_2$ as is clear from the case $w_1 = w_2 \equiv 1$ and (7.32). With $w_2 \equiv 1$ and $w_1 = w_\alpha$, $\alpha > 0$, we thus have under the above assumptions on the parameters,

$$A_{p_1,q_1}^{s_1}(w_\alpha) \hookrightarrow A_{p_2,q_2}^{s_2}.$$

However, for our special weight this can even be extended to values $p_2 < p_1$, as long as $\frac{1}{p_2} < \frac{1}{p_0} = \frac{1}{p_1} + \frac{\alpha}{n}$.

Remark 8.7 We proved even a bit more in [HT94a, 2.4], namely that for $s \in \mathbb{R}$, $0 < p \leq \infty$ ($p < \infty$ in the case of F-spaces), $0 < q \leq \infty$, $\alpha > 0$, and p_0 as above,

$$A_{p,q}^s(w_\alpha) \hookrightarrow \text{weak} - A_{p_0,q}^s, \qquad (8.22)$$

where in the usual definition of $A_{p,q}^s$, Definition 7.1, the spaces L_{p_0} are replaced by $L_{p_0,\infty}$. This is essentially based on Lemma 3.33.

Due to (8.19) we immediately obtain counterparts for other embeddings, e.g.,

$$B_{p_0,p}^{s_0}(w) \hookrightarrow F_{p,q}^s(w) \hookrightarrow B_{p_1,p}^{s_1}(w) \qquad (8.23)$$

for $0 < p_0 < p < p_1 \leq \infty$, $s \in \mathbb{R}$, $s_0 - \frac{n}{p_0} = s - \frac{n}{p} = s_1 - \frac{n}{p_1}$, $0 < q \leq \infty$, and w an admissible weight function.

Proposition 8.8 *Let* $0 < q \leq \infty$, $0 < p < \infty$, *and* $\sigma_p < s < \frac{n}{p}$. *Let* $\alpha > 0$. *Then*

$$\mathfrak{E}_G\left(F_{p,q}^s(w_\alpha)\right) = \left(t^{-\frac{1}{p}+\frac{s}{n}}, p\right) \qquad (8.24)$$

and

$$\mathfrak{E}_G\left(B_{p,q}^s(w_\alpha)\right) = \left(t^{-\frac{1}{p}+\frac{s}{n}}, q\right). \qquad (8.25)$$

P r o o f : Parallel to the proof of Theorem 8.1 and in view of (8.23) it is sufficient to deal with the B-case only as Propositions 3.4(iv) and 4.5 will do the rest. Furthermore, by the characterisation of $B_{p,q}^s$- and $F_{p,q}^s$-spaces via local means, see [Tri92, Sect. 2.4.6, 2.5.3], one easily checks that

$$A_{p,q}^s(w_\alpha) \hookrightarrow A_{p,q}^s, \qquad \alpha > 0. \qquad (8.26)$$

Proposition 3.4(iv) and Theorem 8.1 imply the upper bound,

$$\mathfrak{E}_G^{B_{p,q}^s(w_\alpha)}(t) \leq c \, t^{-\frac{1}{p}+\frac{s}{n}}, \qquad t > 0.$$

Conversely, we found in Step 2 of the proof of Theorem 8.1 that for $j \in \mathbb{N}$,

$$f_j(x) = 2^{j\frac{n}{r}} \psi\left(2^j x\right), \qquad x \in \mathbb{R}^n,$$

are atoms in $B_{p,q}^s$, where ψ is given by (3.36), and the number r, $1 < r < \infty$, denotes again the "foot-point", i.e., $s - \frac{n}{p} = -\frac{n}{r}$. In particular, $\|f_j|B_{p,q}^s\| \leq 1$, and $f_j^*\left(c\, 2^{-jn}\right) \sim 2^{j\frac{n}{r}}$, $j \in \mathbb{N}$. In view of (8.19) and $w_\alpha(x) \leq c$, for $x \in \text{supp } f_j \subset K_1(0)$, $j \in \mathbb{N}$, we thus conclude

$$\left\|f_j|B_{p,q}^s(w_\alpha)\right\| \leq c\left\|w_\alpha f_j|B_{p,q}^s\right\| \leq c'\left\|f_j|B_{p,q}^s\right\| \leq c'',$$

again using "local" characterisation of $B_{p,q}^s$ such as (sub-) atomic decompositions or local means, cf. Theorem 7.8 or [Tri92, Sect. 2.5.3]. Because of

$$\mathcal{E}_{\mathsf{G}}^{B_{p,q}^s(w_\alpha)}\left(c\, 2^{-jn}\right) \geq f_j^*\left(c\, 2^{-jn}\right) \sim 2^{j\frac{n}{r}}, \quad j \in \mathbb{N},$$

this completes the B-result for the envelope function. Moreover, together with (8.26), Theorem 8.1 and Proposition 4.5 this implies $u_{\mathsf{G}}^{B_{p,q}^s(w_\alpha)} \leq q$. It remains to check the converse inequality, that is, we claim that

$$\left(\int_0^\varepsilon \left[t^{\frac{1}{r}}\, f^*(t)\right]^v \frac{dt}{t}\right)^{1/v} \leq c\, \|f|B_{p,q}^s(w_\alpha)\| \tag{8.27}$$

for all $f \in B_{p,q}^s(w_\alpha)$, if, and only if, $v \geq q$. We use a method similar to Step 3 in the proof of Theorem 8.1, recall also our construction in the proof of Proposition 4.10. Let $b = \{b_j\}_{j \in \mathbb{N}}$ be a sequence of non-negative numbers, and let $x^0 \in \mathbb{R}^n$, $|x^0| > 4$, such that $\text{supp } \psi\left(2^j \cdot -x^0\right) \cap \text{supp } \psi\left(2^k \cdot -x^0\right) = \emptyset$ for $j \neq k$, $j, k \in \mathbb{N}_0$, and ψ given by (3.36). Then

$$f_b(x) := \langle x \rangle^{-\alpha} \sum_{j=1}^\infty 2^{j\frac{n}{r}}\, b_j\, \psi\left(2^j x - x^0\right), \quad x \in \mathbb{R}^n, \tag{8.28}$$

is an atomic decomposition in $B_{p,q}^s(w_\alpha)$, as $w_\alpha f \in B_{p,q}^s$, due to (8.19), and $2^{j\frac{n}{r}}\, \psi\left(2^j x - x^0\right)$, $j \in \mathbb{N}$, are $(s,p)_{K,-1}$-atoms, $K \geq 1$ can be chosen arbitrarily large (no moment conditions needed). Assuming $b = \{b_j\}_{j \in \mathbb{N}} \in \ell_q$, Theorem 7.8 together with (8.19) imply,

$$\left\|f_b|B_{p,q}^s(w_\alpha)\right\| \sim \left\|w_\alpha\, f_b|B_{p,q}^s\right\| \leq c\, \|b|\ell_q\|, \tag{8.29}$$

where c is independent of $b \in \ell_q$. On the other hand,

$$f_b^*\left(c2^{-jn}\right) \geq c'\, 2^{j\frac{n}{r}}\, b_j\, w_\alpha\left(2^{-j}x^0\right)^{-1}, \quad j \in \mathbb{N}.$$

For convenience we may assume $b_1 = \cdots = b_{J-1} = 0$, where $J \in \mathbb{N}$ is suitably chosen such that $2^{-J} \leq \varepsilon$, then $w_\alpha \left(2^{-j} x^0 \right) \sim 1$, hence

$$\left(\sum_{j=J} b_j^v \right)^{1/v} \leq c_1 \left(\sum_{j=J}^{\infty} \left[2^{-j \frac{n}{r}} f_b^* \left(c \, 2^{-jn} \right) \right]^v \right)^{1/v}$$

$$\leq c_2 \left(\int_0^\varepsilon \left[t^{\frac{1}{r}} f_b^*(t) \right]^v \frac{dt}{t} \right)^{1/v}$$

$$\leq c_3 \left\| f_b | B_{p,q}^s(w_\alpha) \right\| \leq c_4 \left\| b | \ell_q \right\|, \tag{8.30}$$

thus disproving (8.27) for $v < q$. Now $u_G^{B_{p,q}^s(w_\alpha)} \geq q$ follows immediately thus – in view of our above remarks – completing the proof. $\qquad\square$

Remark 8.9 Note that – as in the case of L_p-spaces, see Theorem 4.7(i) and Proposition 4.13 – the locally regular weight $w_\alpha(x) = \langle x \rangle^\alpha$, $\alpha > 0$ does not change the local singularity behaviour characterised by $\mathfrak{E}_G(X)$, see Theorem 8.1. Again, the result can be transferred without big difficulties to admissible weights w which are bounded from below, $w(x) \geq c$, $x \in \mathbb{R}^n$. We shall see in Proposition 10.21 that the global behaviour takes care of the weight involved.

We conclude this section with the study of spaces $A_{p,q}^s(w^\alpha)$, where the weight is given by (3.52). These spaces are defined in a way completely parallel to Definition 7.1 replacing $L_p(\mathbb{R}^n)$ by $L_p(\mathbb{R}^n, w) = L_p(w)$, where w is a Muckenhoupt weight in the sense of Section 3.4. Spaces of Besov and Triebel-Lizorkin type with weights w belonging to a Muckenhoupt class $\mathcal{A}_r$ have been treated systematically by Bui et al. in [Bui82], [Bui84], and [BPT96], [BPT97]. Later this topic was revived and extended by Rychkov in [Ryc01], including also approaches for locally regular (admissible) weights. Quite recently, the topic was revived in papers of Roudenko [Rou04], [FR04] and Bownik [Bow05], [BH06]. We studied atomic decompositions of spaces $A_{p,q}^s(\mathbb{R}^n, w)$ in [HPxx]. This will enable us to deal with their growth envelope functions. Note that we do not have the counterpart of (8.19) in this case, whereas assertions (8.22) and (8.23) remain valid.

Proposition 8.10 Let $0 < q \leq \infty$, $0 < p < \infty$, $\sigma_p < s < \frac{n}{p}$. Let $0 < \alpha < n + s - \frac{n}{p}$. Then

$$\mathfrak{E}_G \left(F_{p,q}^s(w^\alpha) \right) = \left(t^{-\frac{1}{p} + \frac{s}{n} - \frac{\alpha}{n}}, \, p \right) \tag{8.31}$$

and

$$\mathfrak{E}_G \left(B_{p,q}^s(w^\alpha) \right) = \left(t^{-\frac{1}{p} + \frac{s}{n} - \frac{\alpha}{n}}, \, q \right). \tag{8.32}$$

P r o o f :   In view of (8.23) we may restrict ourselves to the B-setting only. Let again r denote the number that satisfies $1 < r < \infty$, $s - \frac{n}{p} = -\frac{n}{r}$.

Step 1.   We first deal with the envelope functions. Note that our assumptions imply

$$0 < \frac{1}{r_\alpha} := \frac{\alpha}{n} + \frac{1}{r} < 1. \tag{8.33}$$

We apply an embedding result for spaces $B^s_{p,q}(w^\alpha)$ that can easily be verified using its atomic representation; cf. [HPxx], [Bow05]. Then a sequence space argument as in [KLSS06b, Thm. 1] gives the counterpart of (8.22), and, moreover,

$$B^{s_1}_{p_1,q}(w^\alpha) \hookrightarrow B^{s_2}_{p_2,q} \quad \text{if} \quad 0 < \alpha \le \delta = s_1 - \frac{n}{p_1} - s_2 + \frac{n}{p_2}, \tag{8.34}$$

$s_1 > s_2$, $0 < p_1, p_2 \le \infty$, with $\frac{1}{p_2} < \frac{1}{p_1} + \frac{\alpha}{n}$, $0 < q \le \infty$; see also [HS06]. For convenience we only stated the embedding result (8.34) when $q_1 = q_2 = q$; for extensions and further remarks see Remark 11.9 below. Consequently, choosing σ such that $0 < \sigma < s$, and ϱ such that $s - \frac{n}{p} = \sigma - \frac{n}{\varrho} + \alpha$, (8.34) leads to

$$B^s_{p,q}(w^\alpha) \hookrightarrow B^\sigma_{\varrho,q} \tag{8.35}$$

with

$$\sigma - \frac{n}{\varrho} = -\frac{n}{r} - \alpha = -\frac{n}{r_\alpha}.$$

By (8.33) we can apply Theorem 8.1 together with Proposition 3.4(iv) and obtain

$$\mathcal{E}^{B^s_{p,q}(w^\alpha)}_{\mathsf{G}}(t) \le c\, \mathcal{E}^{B^\sigma_{\varrho,q}}_{\mathsf{G}}(t) \le c'\, t^{-\frac{1}{r_\alpha}} = c\, t^{-\frac{1}{r}-\frac{\alpha}{n}}, \quad 0 < t < 1. \tag{8.36}$$

Conversely, observe that

$$f^\alpha_j(x) = 2^{j\frac{n}{r}+j\alpha}\, \psi\left(2^j x\right), \quad x \in \mathbb{R}^n, \tag{8.37}$$

are atoms in $B^s_{p,q}(w^\alpha)$, where ψ is given by (3.36); we refer to the atomic decomposition result in [HPxx], [Bow05], and also to [HS06]. Hence,

$$\left\| f^\alpha_j | B^s_{p,q}(w^\alpha) \right\| \le 1, \quad \text{and} \quad \left(f^\alpha_j\right)^*\left(c\, 2^{-jn}\right) \sim 2^{j\frac{n}{r}+j\alpha}, \quad j \in \mathbb{N}.$$

Thus

$$\mathcal{E}^{B^s_{p,q}(w^\alpha)}_{\mathsf{G}}\left(c\, 2^{-jn}\right) \ge \left(f^\alpha_j\right)^*\left(c\, 2^{-jn}\right) \sim 2^{jn\left(\frac{1}{r}+\frac{\alpha}{n}\right)}, \quad j \in \mathbb{N},$$

leading to the inequality converse to (8.36).

Step 2.   Obviously Step 1, combined with (8.35), Theorem 8.1 and Proposition 4.5 yields

$$u^{B^s_{p,q}(w^\alpha)}_{\mathsf{G}} \le q,$$

and it remains to disprove

$$\left(\int_0^\varepsilon \left[t^{\frac{1}{r}+\frac{\alpha}{n}} f^*(t)\right]^v \frac{dt}{t}\right)^{1/v} \le c \left\|f|B^s_{p,q}(w^\alpha)\right\| \tag{8.38}$$

for some $c > 0$ and all $f \in B^s_{p,q}(w^\alpha)$, if $v < q$. As usual, we construct extremal functions built upon the atoms (8.37), i.e., for a sequence $b = \{b_j\}_{j\in\mathbb{N}}$ of non-negative numbers we consider

$$f^\alpha(x) = \sum_{j=1}^\infty b_j f^\alpha_j(x) = \sum_{j=1}^\infty b_j\, 2^{j\frac{n}{r}+j\alpha}\, \psi\left(2^j x\right), \quad x \in \mathbb{R}^n. \tag{8.39}$$

By the above-mentioned atomic decomposition argument this leads to

$$\left\|f^\alpha|B^s_{p,q}(w^\alpha)\right\| \le c\|b|\ell_q\|, \tag{8.40}$$

and

$$(f^\alpha)^*\left(c\, 2^{-jn}\right) \ge c'\, b_j\, 2^{j\frac{n}{r}+j\alpha}, \quad j \in \mathbb{N}, \tag{8.41}$$

so that – choosing $J \in \mathbb{N}$ with $2^{-J} \le \varepsilon$, and $b_1 = \cdots = b_{J-1} = 0$, for convenience – (8.38) and (8.40) lead to

$$\left(\sum_{j=J} b_j^v\right)^{1/v} \le c_1 \left(\sum_{j=J}^\infty \left[2^{-j\frac{n}{r}-j\alpha}\, (f^\alpha)^*\left(c\, 2^{-jn}\right)\right]^v\right)^{1/v}$$

$$\le c_2 \left(\int_0^\varepsilon \left[t^{\frac{1}{r}+\frac{\alpha}{n}}\, (f^\alpha)^*(t)\right]^v \frac{dt}{t}\right)^{1/v}$$

$$\le c_3 \left\|f^\alpha|B^s_{p,q}(w^\alpha)\right\| \le c_4\|b|\ell_q\|. \tag{8.42}$$

Now $u_G^{B^s_{p,q}(w^\alpha)} \ge q$ follows immediately and – in view of our introductory remark about the F-case – the proof is finished. $\square$

Remark 8.11 We formulated Proposition 8.10 in the "sub-critical" setting parallel to Proposition 8.8, i.e., assuming $s - \frac{n}{p} < 0$. However, inspecting the above proof, we see that this assumption is much stronger than necessary, as – unlike the unweighted sub-critical situation Theorem 8.1 and the w_α-weighted counterpart in Proposition 8.8, where we need for the foot-point r in Figure 12, given by $\frac{1}{r} = \frac{1}{p} - \frac{s}{n}$, that $1 < r < \infty$, – the whole argument in the above weighted setting relies on $1 < r_\alpha < \infty$ using notation (8.33). Thus we can accordingly weaken our assumptions and realise that Proposition 8.10 remains true for $0 < q \le \infty$, $s > 0$, $0 < p < \infty$ with

$$\max\left(0, s - \frac{n}{p}\right) < \alpha < n + s - \frac{n}{p}. \tag{8.43}$$

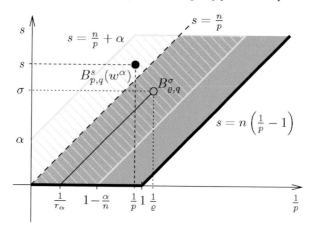

Figure 13

Then, in particular,

$$\mathfrak{E}_\mathsf{G}\left(B^s_{p,q}(w^\alpha)\right) = \left(t^{\frac{s}{n}-\frac{1}{p}-\frac{\alpha}{n}}, q\right). \tag{8.44}$$

This extension of admitted parameters, as indicated in Figure 13, obviously does not apply to weights of type w_α, see Proposition 8.8.

8.2 Growth envelopes in sub-critical borderline cases

We study the situation $s = \sigma_p = n\left(\frac{1}{p}-1\right)_+$ now. Recall that this refers to the thick lines in Figure 12. However, in this situation additional care is needed because not all spaces in question are contained in L_1^{loc}, recall (7.43) and (7.44).

We first consider the *"bottom line"* of the sub-critical strip in Figure 12; that is, where $1 < p < \infty$, and $s = 0$, restricting q according to (7.43), (7.44).

Proposition 8.12 *Let* $1 < p < \infty$.
(i)   *Assume* $0 < q \le 2$. *Then*

$$\mathfrak{E}_\mathsf{G}\left(F^0_{p,q}\right) = \left(t^{-\frac{1}{p}}, p\right). \tag{8.45}$$

(ii)   *Assume* $0 < q \le \min(p,2)$. *Then*

$$\mathfrak{E}_\mathsf{G}\left(B^0_{p,q}\right) = \left(t^{-\frac{1}{p}}, u_\mathsf{G}^{B^0_{p,q}}\right) \qquad with \qquad q \le u_\mathsf{G}^{B^0_{p,q}} \le p. \tag{8.46}$$

In particular,

$$\mathfrak{E}_G\left(B_{p,p}^0\right) = \left(t^{-\frac{1}{p}},\, p\right), \quad 1 < p \le 2.$$

P r o o f :   Step 1.   We use the following embeddings (see (7.32) and (7.7))

$$F_{u,w}^s \hookrightarrow F_{p,q}^0 \hookrightarrow F_{p,2}^0 = L_p, \tag{8.47}$$

where $0 < q \le 2$, $s > 0$, $s - \frac{n}{u} = -\frac{n}{p}$ and $0 < w \le \infty$. Then Theorems 4.7(i), 8.1 and Proposition 3.4(iv) immediately yield

$$\mathcal{E}_G^{F_{p,q}^0}(t) \sim t^{-\frac{1}{p}}.$$

Similarly,

$$B_{u,q}^s \hookrightarrow B_{p,q}^0 \hookrightarrow F_{p,2}^0 = L_p, \tag{8.48}$$

for $0 < q \le \min(p,2)$, $s > 0$, $s - \frac{n}{u} = -\frac{n}{p}$; see (7.32) and (7.27). Hence Theorem 4.7(i) and Proposition 3.4(iv) as well as Theorem 8.1 lead to

$$\mathcal{E}_G^{B_{p,q}^0}(t) \sim t^{-\frac{1}{p}}.$$

Step 2.   Recall that $u_G^{L_p} = p$ by (4.9). Thus Proposition 4.5 together with (8.47) and (8.48) provide

$$u_G^{A_{p,q}^0} \le p,$$

the parameters being as above. We complete the proof for the F-case and proceed by contradiction. Assume $v := u_G^{F_{p,q}^0} < p$; without restriction of generality we may put $v > 1$. Choose a number $s > 0$ such that $\frac{1}{u} := \frac{1}{p} + \frac{s}{n} < \frac{1}{v}$. Now (8.47) and another application of Proposition 4.5 give

$$u_G^{F_{u,w}^s} \le u_G^{F_{p,q}^0} = v.$$

On the other hand we know by Theorem 8.1 that $u_G^{F_{u,w}^s} = u$, thus leading finally to

$$u = u_G^{F_{u,w}^s} \le u_G^{F_{p,q}^0} = v.$$

This, however, contradicts our setting $u > v$. In the same way one proves

$$u_G^{B_{p,q}^0} \ge u_G^{B_{u,q}^s} = q. \qquad \square$$

Remark 8.13 To prove $u_G^{B_{p,q}^0} \ge q$ one could also use the following example, which is an adapted version of [Tri99, Sect. 3.2]. Assume $x^0 \in \mathbb{R}^n$ with $|x^0| > 4$. Put

$$f(x) = \sum_{j=1}^{\infty} b_j\, 2^{j\frac{n}{p}} \left(\psi\left(2^j x\right) - \psi\left(2^j x - x^0\right)\right), \quad x \in \mathbb{R}^n, \tag{8.49}$$

where ψ is the function given by (3.36). Now $2^{j\frac{n}{p}}\left(\psi\left(2^j x\right) - \psi\left(2^j x - x^0\right)\right)$ is an $(0,p)_{K,0}$ - atom in $B_{p,q}^0$ (first moment conditions are necessary), where $K \geq 1$ can be chosen arbitrarily large. Assuming $b = \{b_j\}_{j\in\mathbb{N}} \in \ell_q$, a slightly modified version of [Tri99, (3.13)] (or Theorem 7.8), implies

$$\left\|f|B_{p,q}^0\right\| \leq c\left\|b|\ell_q\right\|, \tag{8.50}$$

where c is independent of $b \in \ell_q$. Moreover, assuming that $b = \{b_j\}_{j\in\mathbb{N}}$ additionally satisfies

$$b_j \geq b_{j+1} \geq d\, 2^{-\frac{n}{p}} b_j > 0, \quad j \in \mathbb{N}, \tag{8.51}$$

for some $d \in \mathbb{R}$, $1 < d < 2^{\frac{n}{p}}$, then

$$f^*\left(2^{-kn}\right) \sim b_k\, 2^{\frac{kn}{p}}, \quad k \in \mathbb{N}. \tag{8.52}$$

Thus, for f given by (8.49) (with the additional assumptions on $b \in \ell_q$), and assuming

$$\left(\int_0^\varepsilon \left[t^{\frac{1}{p}} f^*(t)\right]^v \frac{dt}{t}\right)^{1/v} \leq c\left\|f|B_{p,q}^0\right\| \tag{8.53}$$

for some number v, we obtain by (8.50) and (8.52)

$$\left(\sum_{j=J}^\infty b_j^v\right)^{1/v} \sim \left(\sum_{j=J}^\infty \left[2^{-j\frac{n}{p}} f^*\left(2^{-jn}\right)\right]^v\right)^{1/v} \sim \left(\int_0^\varepsilon \left[t^{\frac{1}{p}} f^*(t)\right]^v \frac{dt}{t}\right)^{1/v}$$

$$\leq c\left\|f|B_{p,q}^0\right\| \leq c'\left(\sum_{j=1}^\infty b_j^q\right)^{1/q}.$$

Now $v \geq q$ follows immediately for all v satisfying (8.53), but this means $u_{\mathsf{G}}^{B_{p,q}^0} \geq q$.

Finally we study the line $s = \sigma_p = n\left(\frac{1}{p} - 1\right)$, where $0 < p \leq 1$, q given by (7.43), (7.44).

Proposition 8.14 *Let* $0 < p \leq 1$ *and* $s = n\left(\frac{1}{p} - 1\right)$.
(i) *Assume* $0 < q \leq \infty$, *and* $0 < q \leq 2$ *if* $p = 1$. *Then*

$$\mathfrak{E}_{\mathsf{G}}\left(F_{p,q}^s\right) = \left(t^{-1}, u_{\mathsf{G}}^{F_{p,q}^s}\right) \quad \text{with} \quad p \leq u_{\mathsf{G}}^{F_{p,q}^s} \leq 1. \tag{8.54}$$

In particular,

$$\mathfrak{E}_{\mathsf{G}}\left(F_{1,q}^0\right) = \left(t^{-1}, 1\right), \quad 0 < q \leq 2.$$

(ii)   *Assume* $0 < q \leq 1$. *Then*

$$\mathfrak{E}_G \left(B^s_{p,q} \right) = \left(t^{-1}, \, u_G^{B^s_{p,q}} \right) \qquad with \qquad q \leq u_G^{B^s_{p,q}} \leq 1. \qquad (8.55)$$

In particular,

$$\mathfrak{E}_G \left(B^s_{p,1} \right) = \left(t^{-1}, \, 1 \right), \quad 0 < p \leq 1, \; s = n \left(\frac{1}{p} - 1 \right).$$

Proof: <u>Step 1.</u>   We have

$$F^s_{p,q} \hookrightarrow F^0_{1,u} \hookrightarrow F^0_{1,2} = h_1 \hookrightarrow L_1, \qquad (8.56)$$

where $0 < u \leq 2$; see (7.10), and [Tri83, Thm. 2.5.8/1, Rem. 2.5.8/4] as well as (7.32) for the embeddings. Thus by Theorem 4.7(i),

$$\mathcal{E}_G^{F^s_{p,q}}(t) \leq c \, \mathcal{E}_G^{F^0_{1,u}}(t) \leq c' \, t^{-1}.$$

Similarly we have

$$B^s_{p,q} \hookrightarrow B^0_{1,q} \hookrightarrow B^0_{1,1} \hookrightarrow L_1, \qquad (8.57)$$

see Proposition 7.11; this implies

$$\mathcal{E}_G^{B^s_{p,q}}(t) \leq c \, \mathcal{E}_G^{B^0_{1,q}}(t) \leq c' \, t^{-1}, \quad 0 < q \leq 1.$$

The corresponding estimates from below can be obtained as in Step 2 of the proof of Theorem 8.1, where r is replaced by 1 now. Moreover, one has to modify the functions f_j given by (8.8) slightly, as now first moment conditions are necessary for atoms in spaces $B^s_{p,q}$ with $s = \sigma_p$, see Theorem 7.8. For $j \in \mathbb{N}$ let the functions f_j be given by

$$f_j(x) := 2^{jn} \left[\psi \left(2^j x \right) - \psi \left(2^j x - x^0 \right) \right], \quad x \in \mathbb{R}^n, \qquad (8.58)$$

and $x^0 \in \mathbb{R}^n$ with, say, $\left| x^0 \right| > 4$; this is a simplified version of Remark 8.13. The function ψ is given by (3.36). Then f_j is an $(s,p)_{K,0}$ - atom in $B^s_{p,q}$, where $K \geq s + 1$. Applying Theorem 7.8 we obtain $\left\| f_j | B^s_{p,q} \right\| \sim 1$ and, parallel to Step 2 of the proof of Proposition 8.1, $f_j^* \left(2^{-jn} \right) \sim 2^{jn}, j \in \mathbb{N}$. Thus again

$$\mathcal{E}_G^{B^s_{p,q}} \left(2^{-jn} \right) \geq f_j^* \left(2^{-jn} \right) \sim 2^{jn}, \quad j \in \mathbb{N}.$$

The rest is a matter of monotonicity, i.e.,

$$\mathcal{E}_G^{B^0_{1,q}}(t) \geq c \, \mathcal{E}_G^{B^s_{p,q}}(t) \geq c' \, t^{-1}, \quad 0 < t < 1.$$

The F-case follows now by elementary embeddings (7.27): choose for a given q, $0 < q \leq 2$, a number u, $0 < u \leq \min(q,1)$. Then $B^0_{1,u} \hookrightarrow F^0_{1,q}$, leading to

$$\mathcal{E}_G^{F^0_{1,q}}(t) \geq c \, \mathcal{E}_G^{B^0_{1,u}}(t) \geq c' \, t^{-1}.$$

When $0 < p < 1$ and $s = n\left(\frac{1}{p} - 1\right)$, we take p_0 such that $0 < p_0 < p$ and put $s_0 = n\left(\frac{1}{p_0} - 1\right)$. Consequently (7.39) implies $B_{p_0,p}^{s_0} \hookrightarrow F_{p,q}^s$, and our just obtained result in the B-case yields

$$\mathcal{E}_{\mathsf{G}}^{F_{p,q}^s}(t) \geq c\, \mathcal{E}_{\mathsf{G}}^{B_{p_0,p}^{s_0}}(t) \geq c'\, t^{-1}.$$

Step 2.   We look for the smallest possible number v such that

$$\left(\int_0^\varepsilon [t\, f^*(t)]^v\, \frac{dt}{t}\right)^{1/v} \leq c\,\|f|A_{p,q}^s\| \tag{8.59}$$

holds for all $f \in A_{p,q}^s$. Clearly (8.56) and (8.57) together with Proposition 4.5 and Theorem 4.7(i) immediately yield $u_{\mathsf{G}}^{A_{p,q}^s} \leq 1$. Furthermore, a slightly modified version of (8.49) serves as an extremal function for the B-setting: with p replaced by 1, as in Remark 8.13, we obtain

$$\|f|B_{p,q}^s\| \leq c\,\|b|\ell_q\|, \qquad \text{and} \qquad f^*\left(2^{-kn}\right) \sim b_k\, 2^{kn}, \quad k \in \mathbb{N},$$

where $b = \{b_j\}_{j\in\mathbb{N}} \in \ell_q$ with the counterpart of (8.51). Hence the B-version of (8.59) implies

$$\left(\sum_{j=J}^\infty b_j^v\right)^{1/v} \sim \left(\sum_{j=J}^\infty [2^{-jn}\, f^*\left(2^{-jn}\right)]^v\right)^{1/v} \sim \left(\int_0^\varepsilon [t\, f^*(t)]^v\, \frac{dt}{t}\right)^{1/v}$$

$$\leq c\,\|f|B_{p,q}^s\| \leq c'\left(\sum_{j=1}^\infty b_j^q\right)^{1/q}$$

for all numbers v satisfying (8.59). Thus $u_{\mathsf{G}}^{B_{p,q}^s} \geq q$, $0 < q \leq 1$; in particular,

$$\mathcal{E}_{\mathsf{G}}\left(B_{p,1}^s\right) = \left(t^{-1}, 1\right), \qquad 0 < p \leq 1, \quad s = \sigma_p.$$

Moreover, this also leads to $u_{\mathsf{G}}^{F_{1,q}^0} \geq 1$, $0 < q \leq 2$, because $B_{p,1}^s \hookrightarrow F_{1,q}^0$ by (7.39). This completes the proof of (8.54) in case of $p = 1$, $s = 0$, i.e.,

$$\mathcal{E}_{\mathsf{G}}\left(F_{1,q}^0\right) = \left(t^{-1}, 1\right), \qquad 0 < q \leq 2.$$

Moreover, let $\sigma > s$ and $\sigma - \frac{n}{r} = s - \frac{n}{p} = \sigma_p$. Then by (7.39), $B_{r,p}^\sigma \hookrightarrow F_{p,q}^s$ and Proposition 4.5 with our above B-result complete the proof in the F-case, too,

$$u_{\mathsf{G}}^{F_{p,q}^s} \geq u_{\mathsf{G}}^{B_{r,p}^\sigma} \geq p. \qquad \square$$

Remark 8.15 Clearly Propositions 8.12 and 8.14 show that the borderline situation, in particular, the determination of the corresponding indices u_G^X, is rather complicated to handle and not yet solved completely (apart from some special cases). Even worse, a reasonable guess as to what the correct outcome could be is also missing. Concerning the *"bottom line"* – referring to Proposition 8.12 – one asks whether B-spaces with $s = 0$ show their "usual" behaviour, i.e., $u_G^{B_{p,q}^0} = q$, independently of the delicate limiting situation, or if they "suffer" from this setting and tend to behave like the F-spaces $u_G^{B_{p,q}^0} = p$ – or something in between. The situation is even more obscure on the line $s = n\left(\frac{1}{p} - 1\right)$, $0 < p \leq 1$.

8.3   Growth envelopes in the critical case

We deal with spaces $A_{p,q}^s$, where $s = \frac{n}{p}$, see Figure 11. Recall that for $0 < p \leq \infty$ (with $p < \infty$ for F-spaces), and $0 < q \leq \infty$,

$$F_{p,q}^{n/p} \hookrightarrow L_\infty \quad \text{if, and only if,} \quad 0 < p \leq 1, \quad 0 < q \leq \infty, \tag{8.60}$$

and

$$B_{p,q}^{n/p} \hookrightarrow L_\infty \quad \text{if, and only if,} \quad 0 < p \leq \infty, \quad 0 < q \leq 1, \tag{8.61}$$

where L_∞ can be replaced by C; see Proposition 7.13. We shall study the remaining cases now.

Theorem 8.16 *Let* $0 < p < \infty$ *and* $0 < q \leq \infty$.

(i)   *Let* $1 < p < \infty$ *and* $\frac{1}{p} + \frac{1}{p'} = 1$, *as usual. Then*

$$\mathfrak{E}_G\left(F_{p,q}^{n/p}\right) = \left(|\log t|^{\frac{1}{p'}}, p\right). \tag{8.62}$$

(ii)   *Let* $1 < q \leq \infty$ *and* $\frac{1}{q} + \frac{1}{q'} = 1$, *as usual. Then*

$$\mathfrak{E}_G\left(B_{p,q}^{n/p}\right) = \left(|\log t|^{\frac{1}{q'}}, q\right). \tag{8.63}$$

P r o o f :   This theorem can be found in [Tri01, Thm. 13.2]; we include here – mainly for completeness and later reference – a slightly different proof as presented in [Har01] and based on [ET99b]. Essential ideas, in particular in Step 3, can be found in [Tri01, Thm. 13.2].

Note that due to (7.39) and Propositions 3.4(iv) and 4.5 it is sufficient to deal with the B-case.

<u>Step 1.</u>   We study the envelope function first and prove

$$\sup_{0<t<\frac{1}{2}} \frac{f^*(t)}{|\log t|^{1/q'}} \leq c \, \left\| f | B_{p,q}^{n/p} \right\| \tag{8.64}$$

for all $f \in B_{p,q}^{n/p}$, where $0 < p < \infty$ and $1 < q \leq \infty$. Assume first $q = \infty$ so that (8.64) reads as

$$\sup_{0<t<\frac{1}{2}} \frac{f^*(t)}{|\log t|} \leq c \, \left\| f | B_{p,\infty}^{n/p} \right\| \tag{8.65}$$

for all $f \in B_{p,\infty}^{n/p}$. Note that by [BS88, (7.22)] one has (at least locally, but this is sufficient for our purpose)

$$\text{bmo} \hookrightarrow L_{\exp,1}, \tag{8.66}$$

see (7.11) for the definition of bmo. On the other hand, Proposition 3.18 yields

$$\mathcal{E}_G^{L_{\exp,a}}(t) \sim |\log t|^a, \quad 0 < t < \frac{1}{2},$$

so by Proposition 3.4(iv) and (8.66) it is sufficient to prove

$$B_{p,q}^{n/p} \hookrightarrow B_{p,\infty}^{n/p} \hookrightarrow \text{bmo}, \quad 0 < p < \infty, \quad 0 < q \leq \infty. \tag{8.67}$$

This can be seen as follows: by (7.10), $h_p = F_{p,2}^0$, $0 < p < \infty$, where h_p are the local Hardy spaces. Then by an application of (7.39),

$$h_1 = F_{1,2}^0 \hookrightarrow B_{r,1}^{-n\left(1-\frac{1}{r}\right)}, \quad 1 < r < \infty,$$

and because of $\text{bmo} = (h_1)'$, see [Gol79b], we arrive at

$$B_{r',\infty}^{n/r'} = \left(B_{r,1}^{-n\left(1-\frac{1}{r}\right)} \right)' \hookrightarrow \left(F_{1,2}^0 \right)' = (h_1)' = \text{bmo}, \quad 1 < r' < \infty, \tag{8.68}$$

where the first identity is covered by [Tri83, Thm. 2.11.2]. By (7.32) we thus have (8.67) for all p, $0 < p < \infty$, and (8.65) is verified. In order to prove (8.64) for $1 < q < \infty$ we use a (nonlinear) real interpolation argument as follows. Let

$$T: \quad f \longmapsto f^{**} \quad \text{and} \quad \mu_a(dt) = \frac{dt}{|\log t|^a}, \tag{8.69}$$

where f^{**} is given by (2.14). Note that one can replace f^* in (8.65) by f^{**}, see [ET99b, Prop. 2.3] or the extended version of Hardy's inequality in [BR80, Thm. 6.4]. Thus (the so modified version of) (8.65) implies that $T : B_{p,\infty}^{n/p} \longrightarrow L_\infty \left(\left(0, \tfrac{1}{2} \right), \mu_1 \right)$. Moreover,

$$
\int\limits_0^{\frac{1}{2}} f^{**}(t)\mathrm{d}t = \int\limits_0^{\frac{1}{2}} \frac{1}{t} \int\limits_0^t f^*(s)\,\mathrm{d}s \;\leq\; \int\limits_0^{\frac{1}{2}} f^*(0)\mathrm{d}t
$$

$$
\leq c\, f^*(0) \;=\; c\, \|f|L_\infty\| \;\leq\; c' \left\| f|B_{p,1}^{n/p} \right\|, \qquad (8.70)
$$

by the monotonicity of f^* and (7.41). Hence $T : B_{p,1}^{n/p} \longrightarrow L_1 \left(\left(0, \tfrac{1}{2} \right), \mu_0 \right)$, with $\mu_0(\mathrm{d}t) = \mathrm{d}t$. We use the replacement of f^* by f^{**} in (8.65) now, because the sub-additivity of f^{**} (2.15) immediately gives the Lipschitz-continuity of T. The nonlinear real interpolation as stated in [Tar72, Thm. 4] leads to

$$
T : \quad \left(B_{p,\infty}^{n/p}, B_{p,1}^{n/p} \right)_{\theta,q} \quad \longrightarrow \quad \left(L_\infty(\mu_1), L_\infty(\mu_0) \right)_{\theta,q} \qquad (8.71)
$$

with $0 < \theta < 1$ and $1 < q < \infty$. Concerning the right-hand side of (8.71) we may proceed by

$$
\left(L_\infty(\mu_1), L_\infty(\mu_0) \right)_{\theta,q} \hookrightarrow \left(L_\infty(\mu_1), L_\infty(\mu_0) \right)_{\theta,\infty} = L_\infty(\mu_{1-\theta}), \qquad (8.72)
$$

see [BL76, Thm. 5.4.1]. On the other hand, for $1 < p < \infty$ and with $\frac{1}{q} = \frac{1-\theta}{\infty} + \frac{\theta}{1} = \theta$ we conclude by [Tri78a, Thm. 2.4.1(b)],

$$
\left(B_{p,\infty}^{n/p}, B_{p,1}^{n/p} \right)_{\theta,q} = B_{p,q}^{n/p}, \quad \theta = \frac{1}{q}, \quad 1 < p < \infty. \qquad (8.73)
$$

Summarising (8.71) – (8.73) results in $T : B_{p,q}^{n/p} \longrightarrow L_\infty(\mu_{1/q'})$, or, equivalently,

$$
\left\| Tf|L_\infty(\mu_{1/q'}) \right\| \;\leq\; c \left\| f|B_{p,q}^{n/p} \right\|, \quad 1 < p < \infty.
$$

In view of (8.69) this yields exactly (8.64) when $1 < p < \infty$. However, the remaining case $0 < p \leq 1$ now simply follows by the monotonicity of B-spaces. In the same way as before (8.64) yields

$$
\mathcal{E}_\mathsf{G}^{B_{p,q}^{n/p}}(t) \leq c \,|\log t|^{\frac{1}{q'}}, \quad 0 < t < \frac{1}{2}.
$$

   Step 2.   We prove $\mathcal{E}_\mathsf{G}^{B_{p,q}^{n/p}}(t) \sim |\log t|^{\frac{1}{q'}}$. We benefit from the construction of extremal functions as presented in [ET99b] by Edmunds and Triebel, recall also Remark 8.13. Let $0 < p < \infty$, $1 \leq q \leq \infty$; put

$$
f_b(x) = \sum_{j=1}^\infty b_j\, \psi \left(2^{j-1} x \right), \quad x \in \mathbb{R}^n, \qquad (8.74)
$$

where ψ is the standard function given by (3.36) and $b = \{b_j\}_{j=1}^{\infty}$ is a sequence of positive, monotonically decreasing numbers with $b \in \ell_q$. Let $J \in \mathbb{N}$ and put $b_j \equiv 1$, $j = 1, \ldots, J$, and $b_j \equiv 0$ for $j > J$; then $\|b_j|\ell_q\| = J^{1/q}$. Denote by f_J the function given by (8.74) with the above-described sequence; then (understanding (8.74) as an atomic decomposition), we have

$$\left\| f_J | B_{p,q}^{n/p} \right\| \leq \|b_j|\ell_q\| \sim J^{\frac{1}{q}}.$$

On the other hand,

$$f_J^*(t) \sim \begin{cases} J & , \quad t \leq 2^{-Jn}, \\ |\log t|, & 2^{-Jn} \leq t \leq 1. \end{cases} \tag{8.75}$$

Hence we have for $0 < p < \infty$, $1 < q \leq \infty$,

$$\mathcal{E}_{\mathsf{G}}^{B_{p,q}^{n/p}}(t) \geq \sup_{J \in \mathbb{N}} J^{-\frac{1}{q}} f_J^*(t) \geq c\, J^{-\frac{1}{q}} |\log t| \sim |\log t|^{\frac{1}{q'}}, \quad t \sim 2^{-Jn}. \tag{8.76}$$

Step 3.   We come to the indices and have thus to prove that

$$\left(\int\limits_0^{\varepsilon} \left[\frac{f^*(t)}{|\log t|^{1/q'+1/v}} \right]^v \frac{dt}{t} \right)^{\frac{1}{v}} \leq c \left\| f | B_{p,q}^{n/p} \right\|, \tag{8.77}$$

is satisfied if, and only if, $v \geq q$. First we prove $u_{\mathsf{G}}^{B_{p,q}^{n/p}} \leq q$. Without restriction of generality we may assume $1 < p < \infty$ now, the extension to $0 < p \leq 1$ being a matter of (7.32) and Proposition 4.5. Moreover, by (8.65) we may also exclude $q = \infty$ here and are thus left to verify

$$\left(\int\limits_0^{\varepsilon} \left[\frac{f^*(t)}{|\log t|} \right]^q \frac{dt}{t} \right)^{\frac{1}{q}} \leq c \left\| f | B_{p,q}^{n/p} \right\|, \qquad 1 < q < \infty. \tag{8.78}$$

Let

$$f(x) = \sum_{j=0}^{\infty} f_j(x) \quad \text{with} \quad f_j(x) = \sum_{m \in \mathbb{Z}^n} \lambda_{jm} a_{jm}(x) \tag{8.79}$$

be an atomic decomposition of $f \in B_{p,q}^{n/p}$ in the sense of Theorem 7.8. Note that we need no moment conditions in our setting. Here $\lambda_{jm} \in \mathbb{C}$, $j \in \mathbb{N}_0$, $m \in \mathbb{Z}^n$, with

$$\Lambda := \|\lambda|b_{pq}\| = \left(\sum_{j=0}^{\infty} \left(\sum_{m \in \mathbb{Z}^n} |\lambda_{jm}|^p \right)^{\frac{q}{p}} \right)^{\frac{1}{q}} < \infty. \tag{8.80}$$

Let χ_{jl} be the characteristic function of the interval $\left[D\,2^{-jn}(l-1), D\,2^{-jn}l\right)$ on $\mathbb{R}_+$, where $D>0$, $j \in \mathbb{N}_0$, and $l \in \mathbb{N}$. Denote by $\left\{\lambda^*_{jl}\right\}_{l\in\mathbb{N}}$ the non-increasing rearrangement of the sequence $\{\lambda_{jm}\}_{m\in\mathbb{Z}^n}$, where $j \in \mathbb{N}_0$ is fixed. If $D>0$ is chosen in an appropriate way, then

$$f^*_j(t) \leq \sum_{l=1}^{\infty} \lambda^*_{jl}\,\chi_{jl}(t), \quad j \in \mathbb{N}_0,$$

see, Example 2.4. Let $t>0$ be such that $D\,2^{-jn}(l-1) \leq t < D\,2^{-jn}l$. Then

$$f^{**}_j(t) = \frac{1}{t}\int_0^t f^*_j(\tau)\mathrm{d}\tau \leq c\,\frac{1}{l}\sum_{k=1}^{l}\lambda^*_{jk} =: c\,\lambda^{**}_{jl}. \tag{8.81}$$

As $1<p<\infty$ and $\left\{\lambda^*_{jk}\right\}_{k=1}^{\infty}$ is a monotonically decreasing sequence, it follows by the sequence version of the Hardy-Littlewood maximal inequality (see, for instance, [CS90, Lemma 1.5.3]) that

$$\sum_{l=1}^{\infty}\left(\lambda^{**}_{jl}\right)^p \sim \sum_{k=1}^{\infty}\left(\lambda^*_{jk}\right)^p = \sum_{m\in\mathbb{Z}^n}|\lambda_{jm}|^p. \tag{8.82}$$

Assume $D\,2^{-(k+1)n} \leq t < D\,2^{-kn}$ for some $k \in \mathbb{N}$. By the sub-additivity (2.15), and (8.79), (8.81) (with $l=1$ and $l=2^{(j-k)n}$, respectively) we arrive at

$$f^{**}(t) \leq \sum_{j=0}^{\infty} f^{**}_j(t) \leq c\sum_{j=0}^{k}\lambda^{**}_{j1} + c\sum_{j=k+1}^{\infty}\lambda^{**}_{j2^{(j-k)n}}.$$

Our assumption $1<q<\infty$ thus implies

$$\int_0^{\frac{1}{2}}\left(\frac{f^*(t)}{|\log t|}\right)^q\frac{\mathrm{d}t}{t} \leq c_1\sum_{k=1}^{\infty}\left(\frac{f^{**}(D\,2^{-kn})}{k}\right)^q$$

$$\leq c_2\sum_{k=1}^{\infty}\left(\frac{1}{k}\sum_{j=0}^{k}\lambda^{**}_{j1}\right)^q + c_2\sum_{k=1}^{\infty}\underbrace{\left(\frac{1}{k}\sum_{j=k+1}^{\infty}\lambda^{**}_{j2^{(j-k)n}}\right)^q}_{=:\Lambda^q_2} \tag{8.83}$$

$$\underbrace{\qquad\qquad}_{=:\Lambda^q_1}$$

The sequence space version of the Hardy-Littlewood maximal inequality yields for the sum Λ^q_1 that

$$\Lambda^q_1 = \sum_{k=1}^{\infty}\left(\frac{1}{k}\sum_{j=0}^{k}\lambda^{**}_{j1}\right)^q \leq c\sum_{j=0}^{\infty}\left(\lambda^{**}_{j1}\right)^q \leq c'\,\Lambda^q, \tag{8.84}$$

where we used (8.82) and (8.80). Put temporarily

$$B_j = \left(\sum_{m \in \mathbb{Z}^n} |\lambda_{jm}|^p \right)^{\frac{1}{p}}, \quad j \in \mathbb{N},$$

for convenience. Then (8.80) reads as $\Lambda^q = \sum_{j=0}^{\infty} B_j^q$. On the other hand, (8.82) and the monotonicity of the sequence $\left\{ \lambda_{jl}^{**} \right\}_{l \in \mathbb{N}}$ – based on the monotonicity of $\left\{ \lambda_{jl}^{*} \right\}_{l \in \mathbb{N}}$ – imply

$$B_j^p \sim \sum_{l=1}^{\infty} \left(\lambda_{jl}^{**} \right)^p = \left(\lambda_{j1}^{**} \right)^p + \sum_{l=1}^{\infty} \sum_{k=2^{(l-1)n}+1}^{2^{ln}} \left(\lambda_{jk}^{**} \right)^p$$

$$\geq c \sum_{l=1}^{\infty} \left(\lambda_{j2^{ln}}^{**} \right)^p 2^{(l-1)n} \geq c \left(\lambda_{j2^{ln}}^{**} \right)^p 2^{(l-1)n}$$

for some $c > 0$ and any $j \in \mathbb{N}_0$, $l \in \mathbb{N}$; that is,

$$\lambda_{j2^{ln}}^{**} \leq C \, 2^{-\frac{nl}{p}} B_j, \quad j \in \mathbb{N}_0, \quad l \in \mathbb{N}.$$

However, this implies

$$\left(\sum_{j=k+1}^{\infty} \lambda_{j2^{(j-k)n}}^{**} \right)^q \leq c \left(\sum_{j=k+1}^{\infty} 2^{-\frac{(j-k)n}{p}} B_j \right)^q \leq c' \sum_{l=0}^{\infty} B_l^q$$

and, for $q > 1$,

$$\Lambda_2^q = \sum_{k=1}^{\infty} \left(\frac{1}{k} \sum_{j=k+1}^{\infty} \lambda_{j2^{(j-k)n}}^{**} \right)^q \leq c \sum_{k=1}^{\infty} \frac{1}{k^q} \sum_{l=0}^{\infty} B_l^q$$

$$\leq c' \sum_{l=0}^{\infty} B_l^q = c' \, \Lambda^q . \tag{8.85}$$

So finally (8.83) together with (8.84) and (8.85) result in

$$\int_0^{\frac{1}{2}} \left(\frac{f^*(t)}{|\log t|} \right)^q \frac{dt}{t} \leq c \, \Lambda . \tag{8.86}$$

Taking the infimum over all representations (8.79) with (8.80) we obtain (8.78), i.e., $u_{\mathrm{G}}^{B_{p,q}^{n/p}} \leq q$, $0 < p < \infty$, $1 < q \leq \infty$.

<u>Step 4.</u>   It remains to show $u_G^{B_{p,q}^{n/p}} \geq q$, i.e., we shall disprove

$$\left(\int_0^\varepsilon \left[\frac{f^*(t)}{|\log t|^{1/q'+1/v}} \right]^v \frac{dt}{t} \right)^{1/v} \leq c \left\| f | B_{p,q}^{n/p} \right\|, \qquad (8.87)$$

for some $c > 0$ and all $f \in B_{p,q}^{n/p}$, whenever $v < q$. We proceed by contradiction; that is, assume $v < q$. Let $J \in \mathbb{N}$ and

$$b_j = j^{-\frac{1}{q}} (\log\langle j \rangle)^{-\frac{1}{v}}, \qquad j = 1, \dots, J;$$

then

$$\left\| f_b | B_{p,q}^{n/p} \right\| \leq c \, \|b|\ell_q\| = c \left(\sum_{j=1}^J \frac{1}{j \, (\log\langle j \rangle)^{q/v}} \right)^{1/q} \leq c_2 < \infty,$$

since $v < q$. Note that c_2 does not depend on $J \in \mathbb{N}$. On the other hand,

$$f_b^*(2^{-kn}) \geq c \sum_{j=1}^k b_j \geq c \, k \, b_k \sim k^{\frac{1}{q'}} (\log\langle k \rangle)^{-\frac{1}{v}}, \qquad k = 1, \dots, J,$$

and so

$$\left(\int_0^\varepsilon \left[\frac{f_b^*(t)}{|\log t|^{1/q'+1/v}} \right]^v \frac{dt}{t} \right)^{\frac{1}{v}} \sim \left(\sum_{k=1}^J \left[\frac{f_b^*(2^{-kn})}{k^{1/q'+1/v}} \right]^v \right)^{\frac{1}{v}} \geq c \left(\sum_{k=1}^J \frac{1}{k \, \log\langle k \rangle} \right)^{\frac{1}{v}}.$$

Now it is clear that the expression on the right-hand side diverges for $J \to \infty$, such that there are functions $f_b \in B_{p,q}^{n/p}$, not satisfying (8.87). This completes the proof.   □

Remark 8.17 In analogy to (8.13) and (8.14) in Remark 8.2 we see that

$$\mathfrak{E}_G \left(L_{\infty,p}(\log L)_{-1} \right) = \left(|\log t|^{\frac{1}{p'}}, p \right) = \mathfrak{E}_G \left(F_{p,q}^{n/p} \right), \qquad (8.88)$$

where $0 < q \leq \infty$ and $1 < p < \infty$; cf. Theorem 4.7(iii) and (8.62). Correspondingly the situation in B-case reads as

$$\mathfrak{E}_G \left(L_{\infty,q}(\log L)_{-1} \right) = \left(|\log t|^{\frac{1}{q'}}, q \right) = \mathfrak{E}_G \left(B_{p,q}^{n/p} \right), \qquad (8.89)$$

where $0 < p < \infty$ and $1 < q \leq \infty$. This follows by Theorem 4.7(iii) and (8.63).

Remark 8.18 Recall Remark 8.4 on the studies of spaces of generalised smoothness by Caetano and Moura; for the critical case their parallel result in [CM04a, Thm. 4.4] reads as

$$\mathfrak{E}_G\left(B_{p,q}^{(n/p,\Psi)}\right) = \left(\Phi_{\infty,q'}(t),\ q\right), \tag{8.90}$$

and

$$\mathfrak{E}_G\left(F_{p,q}^{(n/p,\Psi)}\right) = \left(\Phi_{\infty,p'}(t),\ p\right), \tag{8.91}$$

where Ψ is an admissible function, $0 < p < \infty$, $0 < q \leq \infty$, and $\left(\Psi\left(2^{-j}\right)\right)_{j\in\mathbb{N}} \notin \ell_{q'}$ in the B-case, and $\left(\Psi\left(2^{-j}\right)\right)_{j\in\mathbb{N}} \notin \ell_{p'}$ in the F-case. The auxiliary function $\Phi_{r,u} : (0, 2^{-n}] \to \mathbb{R}$, $0 < r, u \leq \infty$, defined as

$$\Phi_{r,u}(t) := \left(\int_{t^{1/n}}^{1} y^{-\frac{n}{r}u}\ \Psi(y)^{-u}\ \frac{dy}{y} \right)^{1/u},$$

and introduced in [CM04a], has the additional advantage that both the sub-critical and critical result can be summarised using this notation: in our setting we have $\Psi \equiv 1$ and consequently

$$\overline{\Phi}_{r,u}(t) := \left(\int_{t^{1/n}}^{1} y^{-\frac{n}{r}u}\ \frac{dy}{y} \right)^{1/u}, \tag{8.92}$$

leading to the following reformulation of Theorems 8.1 and 8.16: Let $0 < q \leq \infty$, $s > 0$, $1 < r \leq \infty$ and p with $0 < p < \infty$ be such that $s - \frac{n}{p} = -\frac{n}{r}$.

(i) Assume $1 < q \leq \infty$ only if $s = \frac{n}{p}$. Then $\mathfrak{E}_G(B_{p,q}^s) = \left(\overline{\Phi}_{r,q'}(t),\ q\right)$.

(ii) Assume $1 < p < \infty$ only if $s = \frac{n}{p}$. Then $\mathfrak{E}_G(F_{p,q}^s) = \left(\overline{\Phi}_{r,p'}(t),\ p\right)$.

We mention other related work on growth envelopes in the critical case: there is the recent paper on anisotropic results [MNP06], as well as the investigations for spaces of generalised smoothness, [CF06], [CL06]. In the case of spaces with dominating mixed derivatives some envelope results can be found in [KS05], [KS06].

Example 8.19 In view of our Example 7.5 we obtain for $0 < r < \infty$, $0 < u \leq \infty$, $b \in \mathbb{R}$,

$$\Phi_{r,u}^b(t) = \left(\int_{t^{1/n}}^{1} y^{-\frac{n}{r}u}\ (1 + |\log y|)^{-bu}\ \frac{dy}{y} \right)^{1/u} \sim t^{-\frac{1}{r}}\ |\log t|^{-b} \tag{8.93}$$

for $0 < t \leq \frac{1}{2}$, and when $r = \infty$, $b < 1/u$, $0 < u \leq \infty$,

$$\Phi^b_{\infty,u}(t) = \left(\int\limits_{t^{1/n}}^{1} (1 + |\log y|)^{-bu} \frac{dy}{y} \right)^{1/u} \sim |\log t|^{-\left(b - \frac{1}{u}\right)} \qquad (8.94)$$

for $0 < t \leq \frac{1}{2}$; when $b = 1/u > 0$, then

$$\Phi^b_{\infty,u}(t) \sim (\log |\log t|)^{1/u}, \qquad 0 < t \leq \frac{1}{4}. \qquad (8.95)$$

Let $0 < p < \infty$, $0 < q \leq \infty$, $s = \frac{n}{p}$, and $b < 1/q'$ according to Example 7.22, then by (8.94) and (8.90),

$$\mathfrak{E}_G \left(B^{n/p,b}_{p,q} \right) = \left(|\log t|^{\frac{1}{q'} - b}, q \right),$$

whereas for $1 < q \leq \infty$ and $b = 1/q'$, we obtain

$$\mathfrak{E}_G \left(B^{n/p,b}_{p,q} \right) = \left((\log |\log t|)^{1/q'}, q \right).$$

Remark 8.20 Studying spaces on a bounded domain $\Omega \subset \mathbb{R}^n$, say with $|\Omega| < 1$, Theorem 8.16 can be rewritten as

$$F^{n/p}_{p,q}(\Omega) \hookrightarrow L_{\exp,\frac{1}{p'}}(\Omega), \qquad 1 < p < \infty, \quad 0 < q \leq \infty, \qquad (8.96)$$

and

$$B^{n/p}_{p,q}(\Omega) \hookrightarrow L_{\exp,\frac{1}{q'}}(\Omega), \qquad 0 < p < \infty, \quad 1 < q \leq \infty, \qquad (8.97)$$

recall $L_{\infty,\infty}(\log L)_{-a} = L_{\exp,a}$, $a \geq 0$.

Note that (8.96) and (8.97) are nothing else than the classical results mentioned in Remark 3.28, extended to all reasonable cases in the context of B- or F-spaces. Moreover, the history of papers devoted to critical embeddings in the above sense is very long already; we mentioned in Remark 8.3 some of the relevant papers. Additionally we shall refer to Peetre [Pee66], Bennett, Sharpley [BS88, Ch. 4] and Triebel in [Tri93]. and [ET99b, Rem. 2.6] for an extensive discussion of the history of embeddings of that "critical" type.

In view of (7.40) and (7.41) and our preceding comment we may summarise the corresponding results as follows. Recall that we put $\frac{1}{r'} := 0$ when $0 < r < 1$, and $L_{\exp,0} = L_\infty$.

Corollary 8.21 *Let* $\Omega \subset \mathbb{R}^n$ *with* $|\Omega| < 1$. *Let* $0 < p < \infty$ *and* $0 < q \leq \infty$. *Then*

$$F_{p,q}^{n/p}(\Omega) \hookrightarrow L_{\exp,a}(\Omega) \qquad \text{if, and only if,} \qquad a \geq \frac{1}{p'}, \tag{8.98}$$

and

$$B_{p,q}^{n/p}(\Omega) \hookrightarrow L_{\exp,a}(\Omega) \qquad \text{if, and only if,} \qquad a \geq \frac{1}{q'}. \tag{8.99}$$

P r o o f : The cases $0 < p \leq 1$ in (8.98) and $0 < q \leq 1$ in (8.99) are clear, see Proposition 7.13. For the remaining ones, the sufficiency results from (8.96) and (8.97), respectively. Now let $F_{p,q}^{n/p}(\Omega) \hookrightarrow L_{\exp,a}(\Omega)$. We proceed by contradiction; that is, assume $a < \frac{1}{p'}$. Then obviously $1 - a > \frac{1}{p}$ and we may choose a number r with $\frac{1}{p} < \frac{1}{r} < 1 - a$. Consequently,

$$\left(\int_0^{\frac{1}{2}} \left[\frac{f^*(t)}{|\log t|} \right]^r \frac{dt}{t} \right)^{1/r} \leq \sup_{0 < t < \frac{1}{2}} \frac{f^*(t)}{|\log t|^a} \left(\int_0^{\frac{1}{2}} |\log t|^{-(1-a)r} \frac{dt}{t} \right)^{1/r}$$

$$\leq c \sup_{0 < t < \frac{1}{2}} \frac{f^*(t)}{|\log t|^a} \leq c' \, \|f | L_{\exp,a}(\Omega)\|,$$

because $\frac{1}{r} < 1 - a$. Our assumption $F_{p,q}^{n/p}(\Omega) \hookrightarrow L_{\exp,a}(\Omega)$ thus implies

$$\left(\int_0^{\frac{1}{2}} \left[\frac{f^*(t)}{|\log t|} \right]^r \frac{dt}{t} \right)^{1/r} \leq c \left\| f | F_{p,q}^{n/p}(\Omega) \right\|$$

for some $r < p$. This, however, contradicts the sharpness assertion in (8.62). The proof in the B-case is similar. $\qquad\qquad\qquad\square$

Remark 8.22 Note that by (7.40) and (7.41), together with (7.32) and elementary embedding properties of B- and F-spaces it is obvious that $A_{p,q}^s \hookrightarrow L_\infty$ in the super-critical case, see Figure 11. Thus we know that $\mathcal{E}_{\mathsf{G}}^{A_{p,q}^s}(t)$ is bounded in the super-critical case and thus, by our convention, $\mathcal{E}_{\mathsf{G}}(A_{p,q}^s) = (1, \infty)$ when $0 < p \leq \infty$ ($p < \infty$ in the F-case), $s > \frac{n}{p}$, and $0 < q \leq \infty$.

Let $\Omega \subset \mathbb{R}^n$ be bounded, say, with $|\Omega| < 1$, and $\mathrm{bmo}\,(\Omega)$ be the local space of functions of bounded mean oscillation given by (a suitably adapted modification of) (7.11). Note that by (8.66) and Propositions 3.18 and 3.4(iv) we have $\mathfrak{E}_G^{\mathrm{bmo}}(t) \le c \, |\log t|$ for small $t > 0$. Conversely, (8.67) and Theorem 8.16(ii) with $q = \infty$ lead to

$$\mathfrak{E}_G^{\mathrm{bmo}}(t) \ge c \, \mathfrak{E}_G^{B_{p,\infty}^{n/p}}(t) \ge c' \, |\log t| \, .$$

By the same two embedding arguments, now combined with Proposition 4.5, we obtain $u_G^{\mathrm{bmo}} = \infty$, i.e.,

$$\mathfrak{E}_G \left(\mathrm{bmo}\,(\Omega) \right) = (|\log t|, \infty) \, ; \tag{8.100}$$

see also [Tri01, (13.89)]. We finally turn to the situation $p = \infty$, excluded so far and start with the F-spaces. We study the connection between spaces of type $F_{\infty,q}^0$ and bmo, appearing (though secretly hidden) in (8.68) already; recall Remark 7.9. In the critical case $s = 0$, $p = \infty$, one has for $0 < q \le 2$,

$$F_{\infty,q}^0 \hookrightarrow F_{\infty,2}^0 = \mathrm{bmo} \hookrightarrow L_1^{\mathrm{loc}}. \tag{8.101}$$

Conversely, Marschall proved in [Mar95, Lemma 16], that $B_{p,\infty}^{s+n/p} \hookrightarrow F_{\infty,q}^s$ for all $s \in \mathbb{R}$, $0 < p < \infty$, and $0 < q \le \infty$; in particular,

$$B_{p,\infty}^{n/p} \hookrightarrow F_{\infty,q}^0, \quad 0 < p < \infty, \; 0 < q \le \infty \, ; \tag{8.102}$$

the case $q \ge 1$ is already covered by [Mar87, Cor. 4]. Combining (8.100), Theorem 8.16(ii), and (8.101), (8.102) we arrive at

Corollary 8.23 *Let* $0 < q \le 2$, *and* $F_{\infty,q}^0$ *be given by* (7.23). *Then*

$$\mathfrak{E}_G \left(F_{\infty,q}^0 \right) = (|\log t|, \infty) \, .$$

Recall that in the parallel B-case we have by (7.41) that $B_{\infty,q}^0 \hookrightarrow L_\infty$ when $0 < q \le 1$. On the other hand, by (7.44) $B_{\infty,q}^0 \subset L_1^{\mathrm{loc}}$ if, and only if, $0 < q \le 2$. So we are left to consider the case $1 < q \le 2$.

Proposition 8.24 *Let* $1 < q \le 2$. *Then there are positive constants* c_1, c_2 *so that for all small* $t > 0$,

$$c_1 \, |\log t|^{\frac{1}{q'}} \le \mathfrak{E}_G^{B_{\infty,q}^0}(t) \le c_2 \, |\log t| \, . \tag{8.103}$$

P r o o f : Note that (8.63), together with Proposition 3.4(iv) and the elementary embedding $B_{p,q}^{n/p} \hookrightarrow B_{\infty,q}^0$ imply

$$\mathfrak{E}_G^{B_{\infty,q}^0}(t) \ge c \, |\log t|^{\frac{1}{q'}}$$

for small $t > 0$ and all admitted q. On the other hand, by [Tri83, pp. 37, 50, 93] one has $B^0_{\infty,2} \hookrightarrow F^0_{\infty,2} = \mathrm{bmo}$, thus by (8.66) and (4.13) we immediately arrive at

$$\mathcal{E}_{\mathsf{G}}^{B^0_{\infty,q}}(t) \leq c \, |\log t|$$

for any q, $0 < q \leq 2$, and small $t > 0$.   □

Chapter 9

Continuity envelopes $\mathfrak{E}_C$ of function spaces $A_{p,q}^s$

We study continuity envelopes, introduced in Chapters 5 and 6, in the context of function spaces of type $A_{p,q}^s$, where $0 < p \leq \infty$ (with $p < \infty$ in the F-case), $0 < q \leq \infty$, and $\frac{n}{p} \leq s \leq \frac{n}{p} + 1$.

9.1 Continuity envelopes in the super-critical case

We deal with the *super-critical* case of spaces of type $A_{p,q}^s$ as introduced in Figure 11, i.e., $\frac{n}{p} < s \leq \frac{n}{p}+1$. In view of Proposition 7.14 and (7.47) such spaces can be embedded into C. Hence it is reasonable to study their continuity envelope function. On the other hand, when $s > \frac{n}{p}+1$, we may conclude that $A_{p,q}^s$ are continuously embedded in Lip^1, see (7.45) and (7.46), so that by Proposition 5.3(ii) the corresponding continuity envelope functions are bounded and thus of no further interest. We postpone the borderline case $s = \frac{n}{p}+1$ to the next section.

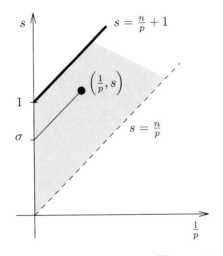

Figure 14

Proposition 9.1 *Let* $0 < q \leq \infty$ *and* $0 < s < 1$. *Then*

$$\mathfrak{E}_C\left(B_{\infty,q}^s\right) = \left(t^{-(1-s)}, q\right). \tag{9.1}$$

P r o o f :   Step 1.   Recall that by (7.3),

$$\|f|B^s_{\infty,q}\| \sim \|f|C\| + \left(\int\limits_0^{\frac{1}{2}} \left[\frac{\omega(f,t)}{t^s} \right]^q \frac{dt}{t} \right)^{1/q}. \tag{9.2}$$

(Note that we consider s as fixed in this context, so that dependences of constants appearing on s do not matter; otherwise one has to deal very carefully with $s \uparrow 1$.) Let $f \in B^s_{\infty,q}$ with $\|f|B^s_{\infty,q}\| \le 1$. Then by (9.2) and Proposition 5.3(i) we obtain for any number τ, $0 < \tau < \frac{1}{2}$,

$$c_1 \frac{\omega(f,\tau)}{\tau} \left(\int\limits_0^\tau t^{(1-s)q-1}\, dt \right)^{\frac{1}{q}} \le c_2 \left(\int\limits_0^\tau \left[\frac{\omega(f,t)}{t^s} \right]^q \frac{dt}{t} \right)^{\frac{1}{q}} \le \|f|B^s_{\infty,q}\| \le 1,$$

hence

$$\frac{\omega(f,\tau)}{\tau} \le c\,\tau^{-(1-s)}.$$

This implies

$$\mathcal{E}^{B^s_{\infty,q}}_C(t) \le c\,t^{-(1-s)}$$

for small values of $t > 0$.

Step 2.   We show the converse inequality; we consider functions f_s given by

$$f_{\sigma,\varkappa}(x) = \sigma^{-\varkappa} \left\{ |x|^{\varkappa+s} \chi_{[0,\sigma)}(|x|) + \psi(x)\chi_{[\sigma,\infty)}(|x|) \right\}, \quad x \in \mathbb{R}^n,$$

with $0 < \sigma < \frac{1}{2}$ and $\varkappa > 0$, and the continuous, non-negative, monotonically decreasing function ψ is chosen so that $f_{\sigma,\varkappa}$ is continuous, $\|f_{\sigma,\varkappa}|C\| \sim \sigma^s < 1$. We obtain $\omega(f_{\sigma,\varkappa},t) \sim \sigma^{-\varkappa}[\min(t,\sigma)]^{\varkappa+s}$ and thus

$$\left(\int\limits_0^{\frac{1}{2}} \left[\frac{\omega(f_{\sigma,\varkappa},t)}{t^s} \right]^q \frac{dt}{t} \right)^{1/q}$$

$$\le c\,\sigma^{-\varkappa} \left\{ \left(\int\limits_0^\sigma t^{\varkappa q-1} dt \right)^{1/q} + \sigma^{\varkappa+s} \left(\int\limits_\sigma^{\frac{1}{2}} t^{-sq-1} dt \right)^{1/q} \right\}$$

$$\le C\,\sigma^{-\varkappa} \left\{ \sigma^\varkappa + \sigma^{\varkappa+s}\sigma^{-s} \right\} \le C',$$

such that $\|f_{\sigma,\varkappa}|B^s_{\infty,q}\| \le c$ for $0 < \sigma < \frac{1}{2}$. On the other hand, for small $t > 0$,

$$\mathcal{E}^{B^s_{\infty,q}}_C(t) \ge \sup_{0<\sigma<\frac{1}{2}} \frac{\omega(f_{\sigma,\varkappa},t)}{t} \ge \frac{\omega(f_{t,\varkappa},t)}{t} \sim t^{-(1-s)}.$$

Step 3.   It remains to determine the smallest number v satisfying (6.2) where $\mu_C(\mathrm{d}t) \sim \frac{\mathrm{d}t}{t}$, that is,

$$\left(\int\limits_0^\varepsilon \left[\frac{\omega(f,t)}{t \cdot t^{-(1-s)}} \right]^v \frac{\mathrm{d}t}{t} \right)^{1/v} \leq c \, \| f | B^s_{\infty,q} \|$$

for $f \in B^s_{\infty,q}$. But in view of (9.2) and Proposition 4.1, $v \geq q$ is obvious. $\square$

We study spaces $A^s_{p,q}$ now belonging to the *"super-critical strip"* (without the borderlines so far), that is, $0 < s - \frac{n}{p} < 1$, $0 < p < \infty$, see Figure 14.

Theorem 9.2 *Let* $0 < p < \infty$, $0 < q \leq \infty$, $\frac{n}{p} < s < \frac{n}{p} + 1$. *Then*

$$\mathfrak{E}_C\left(B^s_{p,q} \right) = \left(t^{-(1+\frac{n}{p}-s)}, \, q \right) \tag{9.3}$$

and

$$\mathfrak{E}_C\left(F^s_{p,q} \right) = \left(t^{-(1+\frac{n}{p}-s)}, \, p \right). \tag{9.4}$$

Proof: Step 1.   We start with the B-setting. Let $\sigma = s - \frac{n}{p}$ be given as in Figure 14, so that $0 < \sigma < 1$. The elementary embedding $B^s_{p,q} \hookrightarrow B^\sigma_{\infty,q}$ and (9.1) imply

$$\mathcal{E}_C^{B^s_{p,q}}(t) \leq c \, t^{-(1-\sigma)}$$

for small $t > 0$. Conversely, let

$$f_j(x) = 2^{-j\sigma} \, \varphi\left(2^j x \right), \quad j \in \mathbb{N}, \tag{9.5}$$

where φ is of the following type. We start with some (non-smooth) function

$$\tilde{\varphi}(x) = \begin{cases} 0 & , \, |x| \geq 1, \\ 1 - |x| & , \, |x| \leq 1, \end{cases} \quad x \in \mathbb{R}^n,$$

such that $\operatorname{supp} \tilde{\varphi}\left(2^j \cdot \right) \subset \{ y \in \mathbb{R}^n : |y| \leq 2^{-j} \}$, $j \in \mathbb{N}$, and

$$\frac{\omega\left(\tilde{\varphi}\left(2^j \cdot \right), t \right)}{t} = 2^j, \quad t \sim 2^{-j}, \quad j \in \mathbb{N}.$$

Now we mollify the function $\tilde{\varphi}$ in a standard way and obtain some smooth descendant φ with similar properties. We realise that f_j given by (9.5) is a $B^s_{p,q}$ - atom (as we do not need moment conditions again). In particular, this means $\| f_j | B^s_{p,q} \| \sim 1$. Moreover,

$$\frac{\omega(f_j, t)}{t} \sim 2^{j(1-\sigma)}, \quad t \sim 2^{-j}, \quad j \in \mathbb{N},$$

so that

$$\mathcal{E}_{\mathsf{C}}^{B_{p,q}^s}\left(2^{-j}\right) \geq c\,\frac{\omega\left(f_j,2^{-j}\right)}{2^{-j}} \geq c'\,2^{j(1-\sigma)}, \quad j \in \mathbb{N},$$

and by standard arguments,

$$\mathcal{E}_{\mathsf{C}}^{B_{p,q}^s}(t) \geq c\,t^{-(1-\sigma)}, \qquad 0 < t < 1,$$

as desired.

Step 2.   We show that

$$\left(\int_0^\varepsilon \left[\frac{\omega(f,t)}{t^\sigma}\right]^v \frac{dt}{t}\right)^{1/v} \leq c\,\|f|B_{p,q}^s\| \tag{9.6}$$

for all $f \in B_{p,q}^s$ if, and only if, $v \geq q$, i.e., $u_{\mathsf{C}}^{B_{p,q}^s} = q$. By $B_{p,q}^s \hookrightarrow B_{\infty,q}^\sigma$ and Propositions 9.1 and 6.4 it follows $u_{\mathsf{C}}^{B_{p,q}^s} \leq u_{\mathsf{C}}^{B_{\infty,q}^\sigma} = q$. The converse inequality needs some more effort. For this purpose we construct extremal functions based on a combination of the functions f_j given by (9.5). Put

$$f(x) := \sum_{j=1}^\infty b_j\,2^{-j\sigma}\,\varphi\left(2^j x - y^j\right), \qquad x \in \mathbb{R}^n, \tag{9.7}$$

where the function φ behaves as described in Step 1 above, and $b_j > 0$, $j \in \mathbb{N}$. Moreover, we can choose $y^j \in \mathbb{R}^n$, $j \in \mathbb{N}$, such that the supports of $\varphi\left(2^j \cdot -y^j\right)$ and $\varphi\left(2^k \cdot -y^k\right)$ are disjoint for $k \neq j$. Note that – in view of our above remarks concerning φ – the functions $g_j(x) = 2^{-j\sigma}\varphi\left(2^j x - y^j\right)$ are $(s,p)_{K,-1}$ - atoms (no moment conditions) in $B_{p,q}^s$, where $K \geq s+1$ can be chosen arbitrarily large, $|D^\gamma g_j(x)| \leq 2^{-j\sigma+j|\gamma|} = 2^{-j(s-n/p)+j|\gamma|}$, $|\gamma| \leq K$. Let $b = \{b_j\}_{j\in\mathbb{N}} \in \ell_q$; then by Theorem 7.8,

$$\|f|B_{p,q}^s\| \leq c\,\|b|\ell_q\|\,. \tag{9.8}$$

In addition, we shall assume now

$$b_{j+1} \leq b_j \leq 2^{1-\sigma}b_{j+1}, \qquad j \in \mathbb{N}. \tag{9.9}$$

Then by construction (9.7) (and our preceding remarks concerning the function φ) we obtain

$$\omega\left(f,2^{-j}\right) \sim 2^{-j}\sup_{k\leq j} b_k\,2^{k(1-\sigma)} = b_j\,2^{-j\sigma}\,.$$

Now it is clear, that (9.9) is needed for this last observation only. Moreover, it is the natural replacement of (8.51) (when dealing with the sub-critical case in terms of growth envelopes). Thus (9.6) implies for those extremal functions f given by (9.7) with (9.8) (in the above-described setting),

$$\left(\sum_{j=J}^{\infty} b_j^v\right)^{1/v} \sim \left(\sum_{j=J}^{\infty}\left[\frac{\omega\left(f,2^{-j}\right)}{2^{-j\sigma}}\right]^v\right)^{1/v} \sim \left(\int_0^\varepsilon \left[\frac{\omega(f,t)}{t^\sigma}\right]^v \frac{dt}{t}\right)^{1/v}$$

$$\leq c\,\|f|B^s_{p,q}\| \leq c' \left(\sum_{j=1}^{\infty} b_j^q\right)^{1/q}$$

and $v \geq q$ is obvious. This finishes the proof of (9.3).

Step 3. It remains to verify (9.4). Let $s_0 > s > \sigma$, $s_0 - \frac{n}{p_0} = s - \frac{n}{p} = \sigma$, then (7.39) implies

$$B^{s_0}_{p_0,p} \hookrightarrow F^s_{p,q} \hookrightarrow B^\sigma_{\infty,p}$$

and Propositions 5.3(iii) and 6.4 together with our just obtained B-results (9.3) and Proposition 9.1 complete the proof.   □

Remark 9.3 Note that with the help of (5.13) and Theorem 8.1 one easily derives $u_C^{B^s_{p,q}} \leq q$ and $u_C^{F^s_{p,q}} \leq p$, respectively, in Theorem 9.2: simply put $\varkappa = \sigma$, $s = \sigma + \frac{n}{p}$, $\frac{1}{r} = \frac{1-\sigma}{n}$, and make use of the lifting property (7.8) for $A^s_{p,q}$-spaces, and also our argument in Step 2 of the proof of Theorem 9.4 below.

Parallel to Remarks 8.2 and 8.17 we mention that Theorem 6.6, Proposition 9.1 and Theorem 9.2 lead to

$$\mathfrak{E}_C\left(B^{\sigma+n/p}_{p,q}\right) = \left(t^{-(1-\sigma)}, q\right) = \mathfrak{E}_C\left(\mathrm{Lip}^{(\sigma,0)}_{\infty,q}\right)$$

for $0 < p \leq \infty$, $0 < q \leq \infty$, $0 < \sigma < 1$, and

$$\mathfrak{E}_C\left(F^{\sigma+n/p}_{p,q}\right) = \left(t^{-(1-\sigma)}, p\right) = \mathfrak{E}_C\left(\mathrm{Lip}^{(\sigma,0)}_{\infty,p}\right)$$

where $0 < p < \infty$, $0 < q \leq \infty$, $0 < \sigma < 1$. Comparison with the situation of spaces of generalised smoothness is postponed to Remark 9.7 below.

9.2   Continuity envelopes in the super-critical borderline case

We come to the borderline case $s = \frac{n}{p} + 1$. Let $0 < p \leq \infty$ (with $p < \infty$ for F-spaces), $0 < q \leq \infty$ and $\alpha \geq 0$. Then, by Proposition 7.15,

$$F_{p,q}^{1+n/p} \hookrightarrow \mathrm{Lip}^{(1,-\alpha)} \qquad \text{if, and only if,} \qquad \alpha \geq \frac{1}{p'}, \qquad (9.10)$$

and

$$B_{p,q}^{1+n/p} \hookrightarrow \mathrm{Lip}^{(1,-\alpha)} \qquad \text{if, and only if,} \qquad \alpha \geq \frac{1}{q'}. \qquad (9.11)$$

In particular,

$$F_{p,q}^{1+n/p} \hookrightarrow \mathrm{Lip}^{1} \quad \text{if, and only if,} \quad 0 < p \leq 1 \quad \text{and} \quad 0 < q \leq \infty, \qquad (9.12)$$

and

$$B_{p,q}^{1+n/p} \hookrightarrow \mathrm{Lip}^{1} \quad \text{if, and only if,} \quad 0 < p \leq \infty \quad \text{and} \quad 0 < q \leq 1, \qquad (9.13)$$

is the (lifted) counterpart of (7.40) and (7.41). Hence, in view of Proposition 5.3(ii) it is clear that spaces given by (9.12) and (9.13), respectively, are of no further interest in our context, because the corresponding envelope functions are bounded. We shall study the remaining cases now. Recall Proposition 5.10; this enables us to give our result in the "borderline" super-critical case when $s = \frac{n}{p} + 1$.

Theorem 9.4 *Let $0 < p < \infty$ and $0 < q \leq \infty$.*

(i)   *Let $1 < p < \infty$ and $\frac{1}{p} + \frac{1}{p'} = 1$, as usual. Then*

$$\mathfrak{E}_{\mathsf{C}}\left(F_{p,q}^{1+n/p}\right) = \left(|\log t|^{\frac{1}{p'}}, p\right). \qquad (9.14)$$

(ii)   *Let $1 < q \leq \infty$ and $\frac{1}{q} + \frac{1}{q'} = 1$, as usual. Then*

$$\mathfrak{E}_{\mathsf{C}}\left(B_{p,q}^{1+n/p}\right) = \left(|\log t|^{\frac{1}{q'}}, q\right). \qquad (9.15)$$

P r o o f :   Step 1.   Clearly, by Propositions 5.3(iii) and Theorem 6.6(ii), as well as (9.10), (9.11), we obtain

$$\mathcal{E}_{\mathsf{C}}^{F_{p,q}^{1+n/p}}(t) \leq c\,|\log t|^{\frac{1}{p'}}, \qquad \mathcal{E}_{\mathsf{C}}^{B_{p,q}^{1+n/p}}(t) \leq c'\,|\log t|^{\frac{1}{q'}}.$$

We have to show the converse inequalities. In view of the construction presented for the proof of Theorem 8.16, we look for some extremal functions in the sense of (8.74). We shall use those constructed by Triebel in [Tri01, (14.15)-(14.19)] in the following way. Let $0 < \delta \le \frac{1}{4}$ and put

$$h(y) := \left\{ \begin{array}{ll} e^{-\frac{1}{\delta^2 - y^2}} & \text{if } |y| < \delta \\ 0 & \text{if } |y| \ge \delta \end{array} \right\}, \qquad y \in \mathbb{R}. \tag{9.16}$$

For $L \in \mathbb{N}_0$ let

$$h_L(y) := h(y) - \sum_{\ell=0}^{L} \mu_\ell \, h^{(\ell)}(y-1),$$

where the numbers $\mu_\ell \in \mathbb{R}$ are chosen so that h_L has moment conditions up to order L,

$$\int_{\mathbb{R}} y^k \, h_L(y) dy = 0, \qquad k = 0, \dots, L.$$

Setting $h^L(y) := \int_{-\infty}^{y} h_L(u) du$, $y \in \mathbb{R}$, we can replace (8.74) by

$$f_b(x) = \sum_{j=1}^{\infty} b_j 2^{-j+1} h^L \left(2^{j-1} x_1 \right) \prod_{k=2}^{n} h \left(2^{j-1} x_k \right), \qquad x = (x_1, \dots, x_n). \tag{9.17}$$

Assume that $b = \{b_j\}_{j=1}^{\infty}$ is a sequence of non-negative monotonically decreasing numbers,

$$b_1 \ge b_2 \ge \dots \ge b_j \ge b_{j+1} \ge \dots \ge 0.$$

Then (9.17) can be interpreted as an atomic decomposition in $A^{1+n/p}_{p,q}$, see [Tri01, Cor. 13.4]. The assertion parallel to the argument presented in the proof of Theorem 8.16 reads as

$$\left(\sum_{j=1}^{\infty} b_j^q \right)^{1/q} \sim \left(\int_0^{\varepsilon} \left[\frac{\omega(f_b, t)}{t \, |\log t|} \right]^q \frac{dt}{t} \right)^{1/q} \sim \left\| f_b | B^{1+n/p}_{p,q} \right\|, \tag{9.18}$$

and a similar expression for F-spaces; cf. [Tri01, (14.18)-(14.19)]. There one also finds the useful estimate

$$\frac{\omega(f_b, t)}{t} \sim \sum_{j=1}^{k} b_j \quad \text{if} \quad t \sim 2^{-k} \quad \text{and} \quad k \in \mathbb{N}. \tag{9.19}$$

We proceed as for Theorem 8.16 and put $b_j \equiv 1$, $j = 1, \dots, J$, and $b_j \equiv 0$ for $j > J$ for some $J \in \mathbb{N}$. Denote by f_J the function given by (9.17) with the above-described sequence; thus – according to (9.18) – $\left\| f_J | B^{1+n/p}_{p,q} \right\| \sim \|b|\ell_q\| = J^{1/q}$. On the other hand one computes by (9.19),

$$\frac{\omega(f_J, 2^{-k})}{2^{-k}} \sim \left\{ \begin{array}{ll} k, & k \le J \\ J, & k > J \end{array} \right. .$$

Hence we obtain for any $J \in \mathbb{N}$,

$$\mathcal{E}_{\mathsf{C}}^{B_{p,q}^{1+n/p}}\left(2^{-J}\right) \geq J^{-\frac{1}{q}} \, \frac{\omega\left(f_J, 2^{-J}\right)}{2^{-J}} \sim J^{\frac{1}{q'}},$$

completing the argument in the B-case. The F-case can be handled in analogy to Theorem 8.16; in particular, (7.39) implies

$$B_{r,p}^{1+n/r} \hookrightarrow F_{p,q}^{1+n/p} \tag{9.20}$$

for $0 < r < p$ leading to $\mathcal{E}_{\mathsf{C}}^{F_{p,q}^{1+n/p}}(t) \sim |\log t|^{\frac{1}{p'}}$ finally.

Step 2. We prove $u_{\mathsf{C}}^{B_{p,q}^{1+n/p}} \leq q$, $u_{\mathsf{C}}^{F_{p,q}^{1+n/p}} \leq p$; that is, we show

$$\left(\int_0^\varepsilon \left[\frac{\omega(f,t)}{t \, |\log t|}\right]^p \frac{dt}{t}\right)^{1/p} \leq c \, \left\|f | F_{p,q}^{1+n/p}\right\| \tag{9.21}$$

when $1 < p < \infty$, $0 < q \leq \infty$, and

$$\left(\int_0^\varepsilon \left[\frac{\omega(f,t)}{t \, |\log t|}\right]^q \frac{dt}{t}\right)^{1/q} \leq c \, \left\|f | B_{p,q}^{1+n/p}\right\| \tag{9.22}$$

when $0 < p < \infty$, $1 < q \leq \infty$. We essentially gain from Proposition 5.10 and our preceding results in Section 8.3 now. Recall (7.9),

$$\left\|f | A_{p,q}^{1+n/p}\right\| \sim \left\|f | A_{p,q}^{n/p}\right\| + \sum_{k=1}^n \left\|\frac{\partial f}{\partial x_k} \,\Big|\, A_{p,q}^{n/p}\right\|. \tag{9.23}$$

We start with the F-case, that is, $1 < p < \infty$. Let first $q < \infty$. Apply (5.12) with $u = 1$, $r = p$; then Theorem 8.16, i.e.,

$$\left(\int_0^{\frac{1}{2}} \left[\frac{g^*(t)}{|\log t|}\right]^p \frac{dt}{t}\right)^{1/p} \leq c \, \left\|g | F_{p,q}^{n/p}\right\|, \tag{9.24}$$

and the F-part of (9.23) yield

$$\left(\int_0^\varepsilon \left[\frac{\omega(f,t)}{t \, |\log t|}\right]^p \frac{dt}{t}\right)^{1/p} \leq c \left(\int_0^\varepsilon \left[\frac{|\nabla f|^*(t)}{|\log t|}\right]^p \frac{dt}{t}\right)^{1/p} \leq c' \, \left\|f | F_{p,q}^{1+n/p}\right\|$$

for all $f \in C^1 \cap F_{p,q}^{1+n/p}$. The rest is done by completion. The same method applies in the B-case when $q < \infty$, now using (5.12) with $r = q$, $u = 1$

(recall our assumption $q > 1$, that is $u = 1 > \frac{1}{q} = \frac{1}{r}$) and the B-counterpart of (9.24),

$$\left(\int\limits_0^\varepsilon \left[\frac{\omega(f,t)}{t|\log t|} \right]^q \frac{dt}{t} \right)^{1/q} \leq c \left(\int\limits_0^\varepsilon \left[\frac{|\nabla f|^*(t)}{|\log t|} \right]^q \frac{dt}{t} \right)^{1/q} \leq c' \left\| f | B^{1+n/p}_{p,q} \right\|.$$

Finally, when $q = \infty$ we shall prove

$$\sup_{0 < t < \varepsilon} \frac{\omega(f,t)}{t\,|\log t|} \leq c \left\| f | B^{1+n/p}_{p,\infty} \right\| \tag{9.25}$$

for all $f \in B^{1+n/p}_{p,\infty}$. (Note that the completion argument does not work here.) Let $\varphi \in \mathcal{S}$ be some standard smooth function compactly supported near the origin, see (7.1). We apply the counterpart of (5.12) (with $r = \infty$) to $f_j = \mathcal{F}^{-1}\varphi\left(2^{-j}\cdot\right)\mathcal{F}f$, $j \in \mathbb{N}$, and obtain (9.25) for any f_j. The right-hand sides can be estimated uniformly with respect to j, thus

$$\sup_{0 < t < \varepsilon} \frac{\omega(f_j,t)}{t\,|\log t|} \leq c \left\| f | B^{1+n/p}_{p,\infty} \right\|$$

for any $j \in \mathbb{N}$. Now f_j converges pointwise to f and (9.25) follows, as well as its counterpart in the F-case.

Step 3. It remains to show the sharpness of $v = p$ (in the F-case) and $\overline{v = q}$ (in the B-case). This works exactly as for Theorem 8.16, now with the extremal functions given by (9.17), (9.18) (and the parallel F-assertion), where we benefit from (9.19) again. $\qquad\qquad\square$

Remark 9.5 Combining Theorem 6.6(ii) and (9.14), (9.15), we arrive at

$$\mathfrak{E}_C\left(\text{Lip}^{(1,-1)}_{\infty,p} \right) = \left(|\log t|^{\frac{1}{p'}} , p \right) = \mathfrak{E}_C\left(F^{1+n/p}_{p,q} \right), \tag{9.26}$$

with $1 < p < \infty$, $0 < q \leq \infty$, and

$$\mathfrak{E}_C\left(\text{Lip}^{(1,-1)}_{\infty,q} \right) = \left(|\log t|^{\frac{1}{q'}} , q \right) = \mathfrak{E}_C\left(B^{1+n/p}_{p,q} \right), \tag{9.27}$$

with $0 < p \leq \infty$, $1 < q \leq \infty$, respectively; see also Proposition 9.6 below. This situation is similar to Remarks 8.2 and 8.17 when dealing with growth envelopes; the corresponding envelopes coincide whereas the underlying spaces do not; cf. [Har00b, Cor. 13, 20] and its extension by Neves [Nev01a]. In addition to the more or less historic references we gave in Remarks 8.3 and 8.20 already, which are partly connected with the super-critical case, too, we shall mention the results by Brézis, Wainger [BW80], by Bourdaud and Lanza de Cristoforis [BL02], approaches based on extrapolation by Edmunds, Krbec

[EK95], Krbec, Schmeisser [KS01], and recently by Neves [Nev01a]. The borderline case was already studied by Zygmund [Zyg45], [Zyg77]. Concerning limiting embeddings we also contributed with some papers [EH99], [EH00], [Har00b].

Proposition 9.6 *Let* $1 < q \le \infty$. *Then*

$$\mathfrak{E}_{\mathsf{C}}\left(B^1_{\infty,q}\right) = \left(|\log t|^{\frac{1}{q'}}, q\right).$$

P r o o f : Step 1.   As before, we gain from Proposition 5.3(iii) and Theorem 6.6(ii) together with (9.11) and obtain

$$\mathcal{E}_{\mathsf{C}}^{B^1_{\infty,q}}(t) \le c \, |\log t|^{\frac{1}{q'}}.$$

As for the lower bound, note that

$$B^{1+n/p}_{p,q} \hookrightarrow B^1_{\infty,q} \; ; \tag{9.28}$$

hence another application of Proposition 5.3(iii) together with our preceding result from Theorem 9.4, in particular Step 1, gives the desired estimate,

$$\mathcal{E}_{\mathsf{C}}^{B^1_{\infty,q}}(t) \sim |\log t|^{\frac{1}{q'}}.$$

Step 2.   We determine the critical index $u_{\mathsf{C}}^{B^1_{\infty,q}}$. We conclude by (9.28), (9.15) and Proposition 6.4 that $u_{\mathsf{C}}^{B^1_{\infty,q}} \ge q$; hence it remains to show

$$\left(\int_0^\varepsilon \left[\frac{\omega(f,t)}{t \, |\log t|}\right]^q \frac{dt}{t}\right)^{1/q} \le c \, \|f|B^1_{\infty,q}\|, \tag{9.29}$$

or, in other words, $B^1_{\infty,q} \hookrightarrow \mathrm{Lip}^{(1,-1)}_{\infty,q}$. This is covered by Proposition 7.15(i) with $p = \infty$, $v = q$. Note that another simple and elegant proof of (9.29) for $1 < q < \infty$ was obtained by Bourdaud, Lanza de Cristoforis in [BL02, Prop. 1], combining Marchaud's inequality (2.22) and Hardy's inequality (see [BS88, Ch. 3, Lemma 3.9]). $\qquad\square$

Remark 9.7 We already mentioned in Remark 8.18 parallel results for spaces of generalised smoothness and introduced the function $\Phi_{r,u} : (0, 2^{-n}] \to \mathbb{R}$, $0 < r, u \le \infty$,

$$\Phi_{r,u}(t) = \left(\int_{t^{1/n}}^1 y^{-\frac{n}{r}u} \, \Psi(y)^{-u} \frac{dy}{y}\right)^{1/u}.$$

In [HM04] and [CH05] we proved that for $0 < p, q \le \infty$ (with $p < \infty$ in the F-case), $0 < \sigma \le 1$, $s = \frac{n}{p} + \sigma$ and Ψ a continuous admissible function, then

(i) with the additional assumption $\left(\Psi\left(2^{-j}\right)^{-1}\right)_{j\in\mathbb{N}} \notin \ell_{q'}$ when $s = \frac{n}{p} + 1$,

$$\mathfrak{E}_C\left(B^{(s,\Psi)}_{p,q}\right) = \left(\Phi_{\frac{n}{1-\sigma}, q'}(t^n), q\right), \qquad (9.30)$$

(ii) and, assuming that $\left(\Psi\left(2^{-j}\right)^{-1}\right)_{j\in\mathbb{N}} \notin \ell_{p'}$ for $s = \frac{n}{p} + 1$,

$$\mathfrak{E}_C\left(F^{(s,\Psi)}_{p,q}\right) = \left(\Phi_{\frac{n}{1-\sigma}, p'}(t^n), p\right). \qquad (9.31)$$

Obviously our above results for $\Psi \equiv 1$ can be rewritten using the function $\overline{\Phi}$ as presented in Remark 8.18. Releasing also the subordinate (dyadic) partition of unity one obtains spaces $A^{\sigma,N}_{p,q}$ of generalised smoothness; their continuity envelopes are studied in [HM06]. Further results on continuity envelopes related to Lorentz-Karamata-Bessel potential spaces $H^s L_{p,q;\Psi}(\mathbb{R}^n)$ with a slowly varying function Ψ were obtained in [GNO05].

Example 9.8 We return to Example 7.5, $s \in \mathbb{R}$, $b \in \mathbb{R}$, $0 < p < \infty$, $0 < q \leq \infty$. In view of Example 7.22 we assume $\frac{n}{p} < s \leq \frac{n}{p} + 1$, referring to the super-critical case. Let $\sigma = s - \frac{n}{p}$. Then (9.30) and (8.93) imply for $0 < \sigma < 1$,

$$\mathfrak{E}_C\left(B^{\sigma+n/p,b}_{p,q}\right) = \left(t^{-(1-\sigma)}|\log t|^{-b}, q\right) = \mathfrak{E}_C\left(\text{Lip}^{(\sigma,b)}_{\infty,q}\right); \qquad (9.32)$$

see Theorem 6.6(iii). When $0 < p, q \leq \infty$, $\sigma = s - \frac{n}{p} = 1$, and $b < 1/q'$ according to Example 7.22, then by (8.94) and (9.30),

$$\mathfrak{E}_C\left(B^{1+n/p,b}_{p,q}\right) = \left(|\log t|^{\frac{1}{q'}-b}, q\right),$$

whereas for $1 < q \leq \infty$ and $b = 1/q'$, we obtain

$$\mathfrak{E}_C\left(B^{1+n/p,b}_{p,q}\right) = \left((\log|\log t|)^{1/q'}, q\right).$$

Corollary 9.9 *Let* $n < p \leq \infty$, $0 < q \leq \infty$, $\alpha > \frac{1}{q}$, *then*

$$\mathfrak{E}_C^{\text{Lip}^{(1,-\alpha)}_{p,q}}(t) \sim t^{-\frac{n}{p}}|\log t|^{\alpha-\frac{1}{q}}, \qquad 0 < t < \frac{1}{2}, \qquad (9.33)$$

(appropriately modified for $p = \infty$*).*

P r o o f : The case $p = \infty$ is covered by Theorem 6.6(ii), where we even obtained a complete envelope result,

$$\mathfrak{E}_C\left(\text{Lip}^{(1,-\alpha)}_{\infty,q}\right) = \left(|\log t|^{\alpha-\frac{1}{q}}, q\right).$$

Assume now $n < p < \infty$, such that $0 < \sigma = 1 - \frac{n}{p} < 1$. We use (7.59) and (7.61),

$$B_{p,\min(q,1)}^{1,-\alpha+\frac{1}{q}} \hookrightarrow \text{Lip}_{p,q}^{(1,-\alpha)} \hookrightarrow B_{p,\infty}^{1,-\alpha+\frac{1}{q}} \qquad (9.34)$$

and Proposition 5.3(iii) and (9.32). $\square$

9.3 Continuity envelopes in the critical case

We return to the *critical* case, already studied in Section 8.3; that is, we consider spaces $A_{p,q}^{n/p}$, see Figure 11. In view of (7.40) and (7.41) (where L_∞ can be replaced by C) we deal with the remaining cases now, not covered by Theorem 8.16 (in terms of growth envelopes $\mathfrak{E}_G$).

Theorem 9.10 *Let* $0 < p \leq \infty$ *and* $0 < q \leq \infty$.

(i)   *Assume* $0 < p \leq 1$. *Then*

$$\mathcal{E}_C^{F_{p,q}^{n/p}}(t) \sim t^{-1}, \qquad 0 < t < 1,$$

and

$$p \leq u_C^{F_{p,q}^{n/p}} \leq \infty. \qquad (9.35)$$

(ii)   *Assume* $0 < q \leq 1$. *Then*

$$\mathcal{E}_C^{B_{p,q}^{n/p}}(t) \sim t^{-1}, \qquad 0 < t < 1,$$

and

$$q \leq u_C^{B_{p,q}^{n/p}} \leq \infty. \qquad (9.36)$$

P r o o f :   Step 1.   We have by Proposition 7.13 that $A_{p,q}^{n/p} \hookrightarrow C$ for the admitted parameters. Thus Theorem 6.6(iv) immediately provides $\mathcal{E}_C^{A_{p,q}^{n/p}}(t) \leq c\, t^{-1}$, $0 < t < 1$. Conversely, note that our construction of the functions f_j in the proof of Proposition 9.2, that is, in (9.5), works for $\sigma = 0$, too. This yields the lower estimate in the B-case (no moment conditions). Moreover, by (7.39) we have

$$B_{r,p}^{n/r} \hookrightarrow F_{p,q}^{n/p}, \qquad (9.37)$$

where $0 < r < p \leq 1$ and $0 < q \leq \infty$. Thus by the preceding observation,

$$\mathcal{E}_C^{F_{p,q}^{n/p}}(t) \geq c\, \mathcal{E}_C^{B_{r,p}^{n/r}}(t) \geq c'\, t^{-1}.$$

Step 2.   We show that $u_C^{B_{p,q}^{n/p}} \geq q$, $u_C^{F_{p,q}^{n/p}} \geq p$ (the upper estimates in (9.35) and (9.36) are clear). Note that the extremal functions (9.7) serve for the proof of the necessity here, too; that is, the construction presented in the proof of Proposition 9.2, in particular Step 2, works also for $\sigma = 0$. Thus we obtain $u_C^{B_{p,q}^{n/p}} \geq q$. Turning to the F-case, let r be such that $0 < r < p \leq 1$, then (9.37) and Proposition 6.4 complete the argument for (9.35). $\square$

Remark 9.11 We briefly discuss the obvious gaps in (9.35) and (9.36). At first glance one might be tempted to prove that $u_C^{B_{p,q}^{n/p}} = q$, $u_C^{F_{p,q}^{n/p}} = p$; however, our methods presented so far fail necessarily in this *limiting case*: assume we would like to show that

$$\left(\int_0^\varepsilon [\omega(f,t)]^q \, \frac{dt}{t} \right)^{1/q} \leq c \left\| f | B_{p,q}^{n/p} \right\| \tag{9.38}$$

holds for all $f \in B_{p,q}^{n/p}$, $0 < p \leq \infty$, $0 < q \leq 1$. The "lifting argument", however, as applied in Step 2 of the proof of Theorem 9.4 quite effectively, cannot be used as our setting now refers to Proposition 5.10(iii), but with $\varkappa = 0$. This is probably not true in general, but at least not covered by Proposition 5.10. Moreover, for several reasons it is not so clear whether the conjecture $u_C^{B_{p,q}^{n/p}} = q$, $u_C^{F_{p,q}^{n/p}} = p$ is the right one for that case.

Chapter 10

Envelope functions $\mathcal{E}_{\mathsf{G}}$ and $\mathcal{E}_{\mathsf{C}}$ revisited

In this section we return to a more general point of view concerning both growth and continuity envelope functions. First we study assertions like Corollaries 3.14, 3.16 in a more general context; then we investigate in what cases the envelope function $\mathcal{E}_{\mathsf{G}}^X$ can be "realised" by some $f \in X$ in the sense that $\mathcal{E}_{\mathsf{G}}^X(t) \sim f^*(t)$, $0 < t < \varepsilon$; similarly for $\mathcal{E}_{\mathsf{C}}^X$. Finally, as mentioned before, we turn to global settings for growth envelope functions, i.e., we consider $\mathcal{E}_{\mathsf{G}}^X(t)$ for $t \to \infty$.

10.1 Spaces on $\mathbb{R}_+$

In this section we insert a short digression on (envelopes of) spaces on $\mathbb{R}_+ = [0, \infty)$ (equipped with $|\cdot|$) first. We pose the question as to whether, say,

$$\mathcal{E}_{\mathsf{G}}^X \in X,$$

and this makes sense only in such spaces. Secondly, we shall study the question as to whether there exists a function in X that realises the growth envelope function $\mathcal{E}_{\mathsf{G}}^X$, i.e.,

$$\mathcal{E}_{\mathsf{G}}^X(t) \sim f^*(t), \quad 0 < t < 1,$$

for some $f \in X$. Such a function will be called an "enveloping function"; this is postponed to Section 10.2 below. Obviously, these topics are closely related.

We first summarise what is already known. In order to simplify the setting further we regard only spaces X on $\Omega = \left[0, \frac{1}{2}\right]$ in the sequel.

Corollary 10.1

(i) Let $0 < p < \infty$, $0 < q \le \infty$, and $a \in \mathbb{R}$. Then

$$\mathcal{E}_{\mathsf{G}}^{L_{p,q}(\log L)_a} \in L_{p,q}(\log L)_a \qquad \textit{if, and only if,} \qquad q = \infty.$$

(ii) *Let $a \geq 0$. Then*

$$\mathcal{E}_G^{L_{\exp,a}} \in L_{\exp,a} \, .$$

(iii) *We have*

$$\mathcal{E}_G^{\mathrm{bmo}} \in \mathrm{bmo} \, .$$

P r o o f : Part (i) follows from Corollaries 3.14, 3.16, (ii) is a consequence of Corollary 3.19, and (8.100) implies (iii), i.e., $\mathcal{E}_G^{\mathrm{bmo}}(t) \sim |\log t|$, $t > 0$ small, and [BS88, Ch. 5, Sect. 7]. $\quad\square$

Also taking the index u_G^X as defined by Definition 4.2 into account, one observes the following peculiarity: whenever $\mathcal{E}_G^X \in X$, then $u_G^X = \infty$; see Theorem 4.7 and (8.100). Thus the following assertion seems natural.

Proposition 10.2 *Let $X \hookrightarrow L_1^{\mathrm{loc}}$ be a function space with $\mathcal{E}_G^X \in X$ and $\mathcal{E}_G^X \neq 0$. Then $u_G^X = \infty$, i.e., $\mathfrak{E}_G(X) = \left(\mathcal{E}_G^X, \infty\right)$, and $\left\|\mathcal{E}_G^X \,|\, X\right\| \geq 1$.*

P r o o f : Step 1. We need not exclude the case $X \hookrightarrow L_\infty$ as Proposition 3.4(iii) together with Remark 4.4 cover that case. So assume now $X \not\hookrightarrow L_\infty$; it is clear by Remark 4.3 that $u_G^X \leq \infty$. It remains to check whether (4.5) is also satisfied for some $v < \infty$. On the other hand, put $f = \mathcal{E}_G^X$ in (4.5); then (4.5) with $v < \infty$ means

$$\left(\int_0^\varepsilon \mu_G(\mathrm{d}t)\right)^{1/v} \leq c \left\|\mathcal{E}_G^X \,|\, X\right\| \tag{10.1}$$

for some $c > 0$. Recall that μ_G is the associated Borel measure with respect to the distribution function $\log \mathcal{E}_G^X$ (or some differentiable function equivalent to $\mathcal{E}_G^X$, respectively), and $\mathcal{E}_G^X \nearrow \infty$ when $t \downarrow 0$. Consequently the left-hand side of (10.1) does not converge and thus our assumption $v < \infty$ leads to a contradiction.

Step 2. We assume $\mathcal{E}_G^X \in X$ and $\mathcal{E}_G^X \neq 0$ in X, i.e., $\left\|\mathcal{E}_G^X|X\right\| > 0$. Then $g := \mathcal{E}_G^X \left\|\mathcal{E}_G^X|X\right\|^{-1}$ is well-defined, $g \in X$ and $\|g|X\| = 1$. Hence

$$\mathcal{E}_G^X(t) \geq g^*(t) = \frac{\left(\mathcal{E}_G^X\right)^*(t)}{\left\|\mathcal{E}_G^X|X\right\|} = \frac{\mathcal{E}_G^X(t)}{\left\|\mathcal{E}_G^X|X\right\|}$$

by Proposition 3.4(i); this gives the result. $\quad\square$

Remark 10.3 Note that in our above examples in Corollary 10.1 we always have

$$\left\| \mathcal{E}_G^X \mid X \right\| \sim 1. \tag{10.2}$$

This is due to the fact that all these spaces are rearrangement-invariant spaces which can be equivalently re-normed to rearrangement-invariant spaces of type $M(X)$, $\|f|M(X)\| = \sup\limits_{t>0} f^{**}(t)\varphi_X(t)$, for the definition of the maximal function $f^{**}(t)$ and the fundamental function $\varphi_X(t)$ we refer to (2.14) and (3.29), respectively; for Lorentz spaces of type $M(X)$ see [BS88, Ch. 2, Sect. 5]. In view of Propositions 3.4(i) and 3.21, together with $(\mathcal{E}_G^X)^{**}(t) \sim (\mathcal{M}\mathcal{E}_G^X)(t) \sim \mathcal{E}_G^X(t)$ in all above-mentioned examples, we immediately obtain (10.2).

Concerning $\mathcal{E}_C^X$ it obviously makes no sense to ask whether $\mathcal{E}_C^X \in X$ with X being a function space on $\Omega = [0, \frac{1}{2}]$, for – apart from the not very interesting case when $\mathcal{E}_C^X$ is bounded, i.e., $X \hookrightarrow \mathrm{Lip}^1$ – we know that $\mathcal{E}_C^X(t) \nearrow \infty$ when $t \downarrow 0$, such that $\mathcal{E}_C^X \notin X$ for all $X \hookrightarrow C$. However, one may replace this by the question

$$t\,\mathcal{E}_C^X(t) \in X \quad ?$$

It is clear by Corollary 5.7(i) that $t\,\mathcal{E}_C^X$ is uniformly bounded, for

$$0 \leq t\,\mathcal{E}_C^X(t) \leq 2 \sup_{\|f|X\|\leq 1} \|f|C\| \leq 2 \left\| \mathrm{id} : X \to C \right\|, \tag{10.3}$$

recall $X \hookrightarrow C$. For convenience, put

$$\mathfrak{e}^X(t) := t\,\mathcal{E}_C^X(t), \quad t \geq 0. \tag{10.4}$$

Corollary 5.7 yields that $\lim\limits_{t\downarrow 0} \mathfrak{e}^X(t) = 0$, $\mathfrak{e}^X$ is monotonically increasing in $t > 0$ and concave, and $\mathfrak{e}^X$ is uniformly bounded, see (10.3). We shall now study the following question: for which spaces X is it true that $\mathfrak{e}^X \in X$. We look for a counterpart of Corollary 10.1 and collect some examples.

Corollary 10.4

(i)   *We have* $\mathfrak{e}^C \in C$.

(ii)   *Let* $0 < a \leq 1$. *Then* $\mathfrak{e}^{\mathrm{Lip}^a} \in \mathrm{Lip}^a$.

(iii) *Let* $0 < q \leq \infty$, $\alpha > \frac{1}{q}$ *(with* $\alpha \geq 0$ *when* $q = \infty$*). Then*

$$\mathfrak{e}^{\mathrm{Lip}_{q,\infty}^{(1,-\alpha)}} \in \mathrm{Lip}_{q,\infty}^{(1,-\alpha)} \qquad \textit{if, and only if,} \qquad q = \infty.$$

(iv) *Let* $0 < a < 1$, $0 < q \le \infty$, $\alpha \in \mathbb{R}$. *Then*

$$\mathfrak{e}^{\mathrm{Lip}^{(a,-\alpha)}_{\infty,q}} \in \mathrm{Lip}^{(a,-\alpha)}_{\infty,q} \qquad \text{if, and only if,} \qquad q = \infty.$$

P r o o f : Proposition 5.15 covers (i), whereas Proposition 5.12 yields for (ii) that $\mathcal{E}_C^{\mathrm{Lip}^a}(t) \sim t^{-(1-a)}$ for $0 < t < 1$, hence $\mathfrak{e}^{\mathrm{Lip}^a}(t) \sim \max(t^a, 1)$ for $t \ge 0$. Then $w\left(\mathfrak{e}^{\mathrm{Lip}^a}, t\right) \sim t^a$ and by (2.25) we obtain $\left\|\mathfrak{e}^{\mathrm{Lip}^a} \,|\, \mathrm{Lip}^a\right\| \sim 1$.

Concerning (iii) we use $\mathcal{E}_C^{\mathrm{Lip}^{(1,-\alpha)}_{\infty,q}}(t) \sim (1 - \log t)^{\alpha - \frac{1}{q}}$ for small $t > 0$, cf. (5.23). Thus $\sup_{t>0} \mathfrak{e}^{\mathrm{Lip}^{(1,-\alpha)}_{\infty,q}}(t) \sim c_{\alpha,q}$ and $w\left(\mathfrak{e}^{\mathrm{Lip}^{(1,-\alpha)}_{\infty,q}}, t\right) \sim t\,(1 - \log t)^{\alpha - \frac{1}{q}}$ for small $t > 0$. The result follows by the definition of $\mathrm{Lip}^{(1,-\alpha)}_{\infty,q}$ now immediately, see (2.35).

Finally, to prove (iv) we use $\mathcal{E}_C^{\mathrm{Lip}^{(a,-\alpha)}_{\infty,q}}(t) \sim t^{-(1-a)}(1 - \log t)^\alpha$ for small $t > 0$, cf. Proposition 5.14. Hence $\sup_{t>0} \mathfrak{e}^{\mathrm{Lip}^{(a,-\alpha)}_{\infty,q}}(t) \sim c_{a,\alpha,q}$ and $w\left(\mathfrak{e}^{\mathrm{Lip}^{(a,-\alpha)}_{\infty,q}}, t\right) \sim t^a\,|\log t|^\alpha$ for small $t > 0$. The definition of $\mathrm{Lip}^{(a,-\alpha)}_{\infty,q}$ completes the proof. $\qquad\square$

Again, Theorem 6.6 now suggests the counterpart of Proposition 10.2 as we observe that whenever $\mathfrak{e}^X \in X$, we also have $u_C^X = \infty$.

Proposition 10.5 *Let* $X \hookrightarrow C$ *be a non-trivial function space with* $\mathfrak{e}^X \in X$. *Then (unless* $\mathfrak{e}^X$ *is a constant) this implies* $u_C^X = \infty$, *i.e.,* $\mathfrak{E}_C(X) = \left(\mathcal{E}_C^X, \infty\right)$, *and* $\left\|\mathfrak{e}^X \,|\, X\right\| \ge 1$.

P r o o f : Step 1. Parallel to the proof of Proposition 10.2 we conclude that the case $X \hookrightarrow \mathrm{Lip}^1$ is covered by Proposition 5.3(ii) together with Remark 6.3; we turn to the remaining cases now. We shall prove that $w\left(\mathfrak{e}^X, t\right) \sim \mathfrak{e}^X(t)$ for small $t > 0$. Afterwards we proceed as in the proof of Proposition 10.2, i.e., put $f = \mathfrak{e}^X$ in (6.2) and argue as in case of $\mathfrak{E}_G(X)$. So it remains to show $w\left(\mathfrak{e}^X, t\right) \sim \mathfrak{e}^X(t)$. Corollary 5.7 implies that $\psi_C = \mathfrak{e}^X$ is increasing, concave near 0, and $\lim_{t \downarrow 0} \mathfrak{e}^X(t) = 0$. But this yields the assertion.

Step 2. Let $\mathfrak{e}^X \in X$. Note that $\mathfrak{e}^X \equiv 0$ would imply that X contains constants only, but this is excluded. Then $g := \mathfrak{e}^X \left\|\mathfrak{e}^X|X\right\|^{-1}$ is well-defined, $g \in X$ and $\|g|X\| = 1$. Hence

$$\mathfrak{e}^X(t) \ge \frac{w\left(\mathfrak{e}^X, t\right)}{\left\|\mathfrak{e}^X|X\right\|}.$$

In view of our above observation $w\left(\mathfrak{e}^X, t\right) \sim \mathfrak{e}^X(t)$ the result is obtained. $\qquad\square$

Remark 10.6 Observe that $\left\| e^{X} \mid X \right\| \sim 1$ for all our examples in Corollary 10.4.

We review our results in Sections 8.1, 8.3, 9.1.

Corollary 10.7 *Let all spaces be defined on* $\Omega = \left[0, \frac{1}{2} \right]$.

(i)   *Let* $0 < q \leq \infty$, $s > 0$, $1 < r < \infty$ *and* $0 < p < \infty$ *be such that* $s - \frac{1}{p} = -\frac{1}{r}$. *Then*

$$\mathcal{E}_{\mathsf{G}}^{B_{p,q}^{s}} \in B_{p,q}^{s} \qquad \textit{if, and only if,} \qquad q = \infty.$$

(ii)   *Let* $1 < q \leq \infty$, *and* $0 < p < \infty$. *Then*

$$\mathcal{E}_{\mathsf{G}}^{B_{p,q}^{1/p}} \in B_{p,q}^{1/p} \qquad \textit{if, and only if,} \qquad q = \infty.$$

(iii)   *Let* $0 < p \leq \infty$, $0 < q \leq \infty$, $0 < \sigma < 1$, *and* $s = \sigma + \frac{1}{p}$. *Then*

$$e^{B_{p,q}^{s}} \in B_{p,q}^{s} \qquad \textit{if, and only if,} \qquad q = \infty.$$

(iv)   *Let* $0 < p \leq \infty$, *and* $1 < q \leq \infty$. *Then*

$$e^{B_{p,q}^{1+1/p}} \in B_{p,q}^{1+1/p} \qquad \textit{if, and only if,} \qquad q = \infty.$$

Proof: By Theorems 8.1 and 8.16 together with Proposition 10.2 it is immediately clear that only B-spaces with $q = \infty$ can satisfy $\mathcal{E}_{\mathsf{G}}^{X} \in X$ as otherwise $u_{\mathsf{G}}^{A_{p,q}^{s}} < \infty$ which contradicts $\mathcal{E}_{\mathsf{G}}^{X} \in X$. So it remains to verify that in the sub-critical case $t^{-1/r} \in B_{p,\infty}^{s}$, $s - \frac{1}{p} = -\frac{1}{r}$ (locally), and $|\log t| \in B_{p,\infty}^{1/p}$, $0 < p < \infty$, referring to the critical case. For $p \geq 1$ a straightforward calculation based on (7.3) was sufficient, but otherwise the atomic characterisation seems to be better adapted: we start with the sub-critical case, i.e., $s - \frac{1}{p} = -\frac{1}{r}$. Let φ be a smooth cut-off function supported near $t = 0$; take, for instance, the standard one from (7.1). Let $\psi_{j}(t) = \varphi(2^{j}t) - \varphi(2^{j+1}t)$, $j \in \mathbb{N}_{0}$, $0 < t < 1$, build a partition of unity; then

$$t^{-\frac{1}{r}} = \varphi(t) t^{-\frac{1}{r}} \sim \sum_{j=0}^{\infty} 2^{-j\left(s - \frac{1}{p}\right)} \underbrace{\psi_{j}(t)\varphi(t) t^{-\frac{1}{r}} 2^{j\left(s-\frac{1}{p}\right)}}_{:= a_{j}(t)}, \ 0 < t < 1, \quad (10.5)$$

where the $a_{j}(t)$, $j \in \mathbb{N}_{0}$, are supported near $\left\{ s \in [0,1] : s \sim 2^{-j} \right\}$, such that $t^{-\frac{1}{r}} \sim 2^{\frac{j}{r}} \sim 2^{-j\left(s - \frac{1}{p}\right)}$, $t \in \operatorname{supp} a_{j}$. Hence (10.5) can be understood as an atomic decomposition of $t^{-\frac{1}{r}}$ (near 0, no moment conditions) with coefficients $\lambda_{j} \equiv 1$, i.e., $\|\lambda | \ell_{\infty}\| = 1$. Theorem 7.8 then implies $t^{-\frac{1}{r}} \in B_{p,\infty}^{s}$. Concerning the critical case we return to our construction (8.74); in particular, with $\varphi(t)$ as above, and $\psi(t) = h(t)$ given by (9.16), i.e., a one-dimensional version of ψ from (3.36), we consider

$$\sum_{j=1}^{\infty} \psi\left(2^{j-1}t\right) \varphi(2t), \tag{10.6}$$

supported near $t = 0$. Then for small $t > 0$,

$$\sum_{j=1}^{\infty} \psi\left(2^{j-1}t\right) \varphi(2t) \sim \sum_{j=1}^{[|\log t|]} 1 \sim |\log t|,$$

i.e., (10.6) can be interpreted as an atomic decomposition for $|\log t|$ near 0 (no moment conditions) with $\lambda_j \sim 1$ and thus $\|\lambda|\ell_\infty\| \sim 1$. Consequently $|\log t| \in B_{p,\infty}^{1/p}$, $0 < p < \infty$ (locally).

Concerning (iii), (iv), Theorems 9.2, 9.4 imply that only B-spaces with $q = \infty$ can satisfy $e^X \in X$, see Proposition 10.5. So we have to show that $t^\sigma \in B_{p,\infty}^{\sigma+1/p}$ for $0 < \sigma < 1$, $0 < p \le \infty$ (at least locally), and $t|\log t| \in B_{p,\infty}^{1+1/p}$, $0 < p \le \infty$. For the super-critical case (iii) we proceed parallel to the sub-critical one in (i), where (10.5) is now replaced by

$$t^\sigma = \varphi(t)\, t^\sigma \sim \sum_{j=0}^{\infty} 2^{-j\left(\sigma+\frac{1}{p}-\frac{1}{p}\right)} \psi_j(t)\, \varphi(t)\, t^\sigma\, 2^{j\sigma}, \quad 0 < t < 1, \tag{10.7}$$

the rest is similar. Concerning (iv) we return to the extremal functions f_b as constructed by Triebel in [Tri01, (14.15)-(14.19)]; see also (9.18). Put $b_j \equiv 1$; then this is essentially the integrated version of (10.6),

$$\sum_{j=1}^{\infty} 2^{-j+1} \Psi\left(2^{j-1}t\right) \varphi(2t), \qquad \Psi(z) = \int_{-\infty}^{z} \psi(u)\mathrm{d}u, \tag{10.8}$$

where $\psi(t)$, $\varphi(t)$ are as above; note that we need no moment conditions. One checks that

$$\sum_{j=1}^{\infty} 2^{-j+1} \Psi\left(2^{j-1}t\right) \varphi(2t) \sim t\,|\log t|, \quad 0 < t < \frac{1}{2},$$

and (10.8) can be understood as the atomic decomposition of $t\,|\log t|$ (near 0). Now (9.18) and the particular choice of the sequence $b \in \ell_\infty$ imply $t\,|\log t| \in B_{p,\infty}^{1+1/p}$, $0 < p \le \infty$. $\qquad \square$

Remark 10.8 Triebel studied a related question in [Tri01, Sect. 17.1], asking under what conditions there are functions $f \in A_{p,q}^s$ such that $f^*(t)$

or $\frac{\omega(f,t)}{t}$ are equivalent to the corresponding growth or continuity envelope functions. By the same arguments as above only B-spaces with $q = \infty$ are left to consider; Triebel applies these outcomes showing that certain Green's functions (of $(\mathrm{id} - \Delta)^{-\frac{n}{2}}$ for the n-dimensional critical case, for instance) generate the corresponding envelope functions.

10.2   Enveloping functions

As already announced we deal with functions $f \in X$ now that (locally) realise the growth or continuity envelope function, $\mathcal{E}_G^X$ and $\mathcal{E}_C^X$, respectively. This idea originates from discussions with H. Triebel. All spaces are defined on $\mathbb{R}^n$, equipped with the Lebesgue measure, unless otherwise stated.

Definition 10.9 *Let* $\varepsilon > 0$.

(i)   *Let* $X \hookrightarrow L_1^{\mathrm{loc}}$ *be a quasi-normed function space with growth envelope function* $\mathcal{E}_G^X$. *Then* $f_G \in X$, $\|f_G|X\| \leq 1$, *is called* (growth) enveloping *function in* X, *if*

$$\mathcal{E}_G^X(t) \sim f_G^*(t), \quad 0 < t < \varepsilon.$$

(ii)   *Let* $X \hookrightarrow C$ *be a quasi-normed function space with continuity envelope function* $\mathcal{E}_C^X$. *Then* $f_C \in X$, $\|f_C|X\| \leq 1$, *is called a* (continuity) enveloping *function in* X, *if*

$$\mathcal{E}_C^X(t) \sim \frac{\omega(f_C,t)}{t}, \quad 0 < t < \varepsilon.$$

Of course, the particular growth/continuity enveloping function will depend on the function space X; for lucidity we shall but omit the additional index if possible. The number $\varepsilon > 0$ is arbitrary, but meant to be small. For convenience, we may think of $\varepsilon < 1$ in general, and $\varepsilon < \frac{1}{2}$ when log-terms are involved, or $\varepsilon < \varepsilon_0(a, \alpha)$ in Corollary 10.13 below, but this is not important. At the moment we are only interested in *local* enveloping functions – and will thus also omit this special notation.

Remark 10.10 By (obvious counterparts of) Propositions 10.2 and 10.5 there cannot exist enveloping functions in spaces X with $u_G^X < \infty$ or $u_C^X < \infty$, respectively. This excludes, in particular, spaces of type $F_{p,q}^s$. We shall see below, that $u_G^X = \infty$ or $u_C^X = \infty$ is necessary, but not sufficient for the existence of associate enveloping functions in X. In continuation of Proposition 10.2 and Remark 10.3 let us mention that in all cases when X possesses an enveloping function, then $\left\|\mathcal{E}_G^X|X\right\| \sim 1$.

Corollary 10.11   *Let $\varepsilon > 0$ be small and all spaces be defined on $\Omega \subseteq \mathbb{R}^n$.*

(i)   *Let $0 < p < \infty$, and $a \in \mathbb{R}$. Then*

$$f_{\mathsf{G}}(x) = |x|^{-\frac{n}{p}} |\log|x||^{-a} \chi_{K_\varepsilon(0)}(x)$$

   is a growth enveloping function in $L_{p,\infty}(\log L)_a$.

(ii)   *Let $a \geq 0$, and $\Omega = K_1(0)$. Then*

$$f_{\mathsf{G}}(x) = |\log|x||^a \chi_{K_\varepsilon(0)}(x)$$

   is a growth enveloping function in $L_{\exp,a}$.

(iii) *Assume $\Omega = K_1(0)$, and let φ be given by (7.1). Then*

$$f_{\mathsf{G}}(x) = |\log|x|| \varphi(x)$$

   is a growth enveloping function of bmo.

(iv)   *Let $0 < p < \infty$, and $\sigma_p < s < \frac{n}{p}$. Let φ be given by (7.1). Then*

$$f_{\mathsf{G}}(x) = |x|^{s-\frac{n}{p}} \varphi(x)$$

   is a growth enveloping function in $B^s_{p,\infty}$.

(v)   *Let $0 < p < \infty$, and let the function φ be given by (7.1). Then*

$$f_{\mathsf{G}}(x) = |\log|x|| \varphi(2x)$$

   is a growth enveloping function in $B^{n/p}_{p,\infty}$.

P r o o f :   In view of Lemma 3.10(ii), in particular, (3.14) and (3.15) with $s = 1$, $\varkappa = \frac{1}{p}$, and $r = p$, part (i) follows from Proposition 3.15. In the same way Lemma 3.10(iv) with $s = 1$ and Proposition 3.18 imply (ii). The local version of (7.11) together with (8.100) lead to (iii). Concerning (iv) and (v) we stress similar arguments as for the proof of Corollary 10.7(iii), (iv): let φ be given by (7.1), and

$$\psi_j(x) = \varphi\left(2^j x\right) - \varphi\left(2^{j+1}x\right), \quad x \in \mathbb{R}^n, \quad j \in \mathbb{N}_0, \tag{10.9}$$

build a partition of unity in $K_1(0)$. Then for $x \in K_1(0)$, $x \neq 0$,

$$|x|^{s-\frac{n}{p}} = |x|^{s-\frac{n}{p}} \varphi(x) \sim \sum_{j=0}^{\infty} 2^{-j\left(s-\frac{n}{p}\right)} \underbrace{\psi_j(x)\varphi(x)}_{:= a_{j0}(x)}, \tag{10.10}$$

where the $a_{j0}(x)$, $j \in \mathbb{N}_0$, are supported in cQ_{j0}. Hence (10.10) can be seen as an atomic decomposition of $|x|^{s-\frac{n}{p}} \varphi(x)$, (no moment conditions) with coefficients $\lambda_{j0} \equiv 1$, i.e., $\|f_\mathsf{G}|B_{p,\infty}^s\| \le c\|\lambda|\ell_\infty\| \le c'$. Moreover, $f_\mathsf{G}^*(t) \sim t^{\frac{s}{n}-\frac{1}{p}}$, $0 < t < 1$, so Theorem 8.1 completes the argument for (iv). In the critical case, let ψ be given by (3.36); then we claim that

$$f_\mathsf{G}(x) \sim \sum_{j=1}^{\infty} \psi\left(2^{j-1}x\right) \varphi(2x), \tag{10.11}$$

near 0, because for small x,

$$\sum_{j=1}^{\infty} \psi\left(2^{j-1}x\right) \varphi(2x) \sim \sum_{j=1}^{[|\log|x||]} 1 \sim |\log|x||.$$

Interpreting this as an atomic decomposition again (no moment conditions), we obtain $\left\|f_\mathsf{G}|B_{p,\infty}^{n/p}\right\| \le c\|\lambda|\ell_\infty\| \le c'$. The rest is obvious,

$$f_\mathsf{G}^*(t) \sim |\log t| \sim \mathcal{E}_\mathsf{G}^{B_{p,\infty}^{n/p}}(t), \quad 0 < t < \varepsilon,$$

recall Theorem 8.16(ii). $\qquad\square$

We come to continuity enveloping functions.

Remark 10.12 Obviously the first candidate to deal with is $X = C$, the space of bounded uniformly continuous functions. In view of Proposition 5.15 the question can thus be reformulated to find some continuous function f_C such that $\omega(f_\mathsf{C}, t) \sim 1$ for all small $t > 0$. This, however, is impossible for a continuous function, see also the (pointwise) construction in (5.26) for $n \to \infty$. Thus C possesses no continuity enveloping function, though $u_\mathsf{C}^C = \infty$, i.e., this also yields the insufficiency of this condition for the existence of f_C.

Corollary 10.13   *Let $\varepsilon > 0$, and all spaces be defined on $\Omega \subseteq \mathbb{R}^n$.*
(i)   *Let $0 < a \le 1$, $b \in \mathbb{R}$ (with $b \ge 0$ if $a = 1$), and φ be given by (7.1). Then*

$$f_\mathsf{C}(x) = |x|^a \, |\log|x||^b \, \varphi(2x)$$

is a continuity enveloping function in $\mathrm{Lip}_{\infty,\infty}^{(a,-b)}$, in particular,

$$f_\mathsf{C}(x) = |x|^a \, \varphi(x)$$

is associated to Lip^a.

(ii)   *Let $0 < p \le \infty$, $\frac{n}{p} < s < \frac{n}{p} + 1$. Let φ be given by (7.1). Then*

$$f_\mathsf{C}(x) = |x|^{s-\frac{n}{p}} \, \varphi(x)$$

is a continuity enveloping function in $B_{p,\infty}^s$.

(iii) *Let $0 < p \leq \infty$, and φ be given by (7.1). Then*

$$f_{\mathsf{C}}(x) = |x| \, |\log |x|| \; \varphi(2x)$$

is a continuity enveloping function in $B_{p,\infty}^{1+n/p}$.

P r o o f : For convenience we shall assume that $\varepsilon < \min\left(\frac{1}{2}, e^{-b/a}\right)$ in (i). Then clearly

$$\omega\left(f_{\mathsf{C}}, t\right) = t^a \, |\log t|^b, \quad 0 < t < \varepsilon,$$

leads to $\left\| f_{\mathsf{C}} | \mathrm{Lip}_{\infty,\infty}^{(a,-b)} \right\| \leq c$, and

$$\mathcal{E}_{\mathsf{C}}^{\mathrm{Lip}_{\infty,\infty}^{(a,-b)}}(t) \sim \frac{\omega\left(f_{\mathsf{C}}, t\right)}{t}, \quad 0 < t < \varepsilon,$$

recall Propositions 5.12, 5.13 and 5.14. For (ii) we argue similar to Corollary 10.11 by atomic decomposition. We already know that $\left\| f_{\mathsf{C}} | B_{p,\infty}^s \right\| \leq c$ by (10.10) and the construction gives $\omega\left(f_{\mathsf{C}}, t\right) = t^{s-\frac{n}{p}}, 0 < t < 1$. In view of Proposition 9.1 and Theorem 9.2 this concludes the proof of (ii). Finally, in (iii) it remains to prove $\left\| f_{\mathsf{C}} | B_{p,\infty}^{1+n/p} \right\| \leq c$, as the rest is obvious by construction, Theorem 9.4 and Proposition 9.6,

$$\mathcal{E}_{\mathsf{C}}^{B_{p,\infty}^{1+n/p}}(t) \sim |\log t| \sim \frac{\omega\left(f_{\mathsf{C}}, t\right)}{t}, \quad 0 < t < \varepsilon.$$

We "integrate" the atomic decomposition used in Corollary 10.11(v) and proceed just as for Theorem 9.4, in particular (9.17): let h be given by (9.16), and

$$\Psi(y) = \prod_{j=2}^{n} h(y_j) \int_{-\infty}^{y_1} h(u)\mathrm{d}u, \quad y = (y_1, \ldots, y_n) \in \mathbb{R}^n. \tag{10.12}$$

Then

$$\sum_{j=1}^{\infty} 2^{-j+1} \Psi\left(2^{j-1}x\right) \varphi(2x) \sim |x| \, |\log |x||, \quad x \in K_\varepsilon(0), \tag{10.13}$$

can be understood as the atomic decomposition of f_{C} near 0 (no moment conditions needed), see [Tri01, Cor. 13.4]. Hence $\left\| f_{\mathsf{C}} | B_{p,\infty}^{1+n/p} \right\| \leq c$ as desired. □

10.3   Global versus local assertions

As already mentioned several times, we have studied local assertions so far, meaning the behaviour of $\mathcal{E}_{\mathsf{G}}^X(t)$ for small $t > 0$. We shall see also that global assertions are of some interest, in particular in limiting or weighted situations. Some results are contained in [Harxx]. First we recall what is already known from Section 3.2 in this context. For convenience, we shall restrict ourselves to the situation of $\mathbb{R}^n$ equipped with the Lebesgue measure always, $[\mathbb{R}^n, |\cdot|]$ as underlying space.

Corollary 10.14 *Let $0 < p, q \le \infty$ (with $q = \infty$ when $p = \infty$).*
(i)   *Then*

$$\mathcal{E}_G^{L_{p,q}}(t) \sim t^{-\frac{1}{p}}, \qquad t \to \infty. \tag{10.14}$$

(ii)   *Assume $a \in \mathbb{R}$, then*

$$\mathcal{E}_G^{L_{p,q}(\log L)_a}(t) \sim t^{-\frac{1}{p}}(1 + |\log t|)^{-a}, \qquad t \to \infty. \tag{10.15}$$

P r o o f :   This is only a reformulation of (3.22) and (3.24).   □

Proposition 10.15 *Let $1 \le p < \infty$ and $k \in \mathbb{N}_0$. Then*

$$\mathcal{E}_G^{W_p^k}(t) \sim t^{-\frac{1}{p}}, \quad t \to \infty.$$

P r o o f :   By definition (2.37), $W_p^k(\mathbb{R}^n) \hookrightarrow L_p(\mathbb{R}^n)$, thus (10.14) and Proposition 3.4(iv) yield

$$\mathcal{E}_G^{W_p^k}(t) \le c\, t^{-\frac{1}{p}}, \quad t \to \infty.$$

Conversely, we return to our construction in Section 3.4, and modify the functions f_R from (3.35) slightly by

$$f^R(x) = R^{-\frac{n}{p}}\, \psi\left(R^{-1}x\right), \quad x \in \mathbb{R}^n,$$

where $\psi(x)$ is given by (3.36). Similar to the proof of Proposition 3.25 we are led to

$$\left\|f^R | W_p^k\right\| \le \left(\sum_{|\alpha| \le k} R^{-|\alpha|p}\, \|D^\alpha \psi | L_p\|^p\right)^{1/p} \le \|\psi | W_p^k\|,$$

now assuming $R > 1$. Let t be large and choose $R_0 = R_0(t) = dt^{1/n}$ such that $R_0 > (t/|\omega_n|)^{1/n} > 1$, i.e., $R_0^{-n}t < |\omega_n|$ for appropriate $d > 0$. Hence (2.9) and (3.39) imply

$$\mathcal{E}_G^{W_p^k}(t) \ge c_1 \sup_{R>1} \left(f^R\right)^*(t) \ge c_1 \left(f^{R_0}\right)^*(t) \ge c_2 R_0^{-\frac{n}{p}} \ge c_3\, t^{-\frac{1}{p}},$$

the proof is finished.   □

Remark 10.16 If we compare Propositions 3.25 and 10.15 we see that the additional smoothness assumption of W_p^k unlike L_p, that is, $k \in \mathbb{N}_0$, is well-reflected in the local singularity behaviour, $\mathcal{E}_{\mathsf{G}}^{W_p^k}(t)$, $0 < t < 1$, whereas globally the spaces all share the same integrability, $\mathcal{E}_{\mathsf{G}}^{W_p^k}(t) \sim \mathcal{E}_{\mathsf{G}}^{L_p}(t)$, $t \to \infty$.

Proposition 10.17 *Let $1 < p < \infty$, $0 < \alpha < \frac{n}{p'}$. Then*

$$\mathcal{E}_{\mathsf{G}}^{L_p(w_\alpha)}(t) \sim \mathcal{E}_{\mathsf{G}}^{L_p(w^\alpha)}(t) \sim t^{-\frac{\alpha}{n} - \frac{1}{p}}, \quad t \to \infty. \tag{10.16}$$

P r o o f : In case of w^α the result is already covered by Proposition 3.35. Consider now w_α, then Lemma 3.33 and (10.14) imply $\mathcal{E}_{\mathsf{G}}^{L_p(w_\alpha)}(t) \leq t^{-\left(\frac{\alpha}{n} + \frac{1}{p}\right)}$ for $t \to \infty$. Conversely, inspecting the proof of Proposition 3.30 we observe that with extremal functions f_s given by (3.11) with $r = p_0$, $A_s = K_{cs^{1/n}}(0)$, with $c > 0$ such that $\mu(A_s) = |K_{cs^{1/n}}(0)| = s$, we get for large s,

$$\|f_s|L_p(w_\alpha)\| = s^{-\frac{1}{p_0}} \left(\int_{A_s} w_\alpha(x)^p dx \right)^{\frac{1}{p}} = s^{-\frac{1}{p_0}} \left(\int_{A_s} \left(1 + c's^{2/n} \right)^{\frac{\alpha p}{2}} dx \right)^{\frac{1}{p}}$$

$$\leq c'' \, s^{-\frac{1}{p_0}} s^{\frac{\alpha}{n}} |A_s|^{\frac{1}{p}} \quad \leq \quad C.$$

Here p_0 is given by $\frac{1}{p_0} = \frac{1}{p} + \frac{\alpha}{n}$. On the other hand, $f_s^*(t) = s^{-\frac{1}{p_0}} \chi_{[0,s)}(t)$, resulting in

$$\mathcal{E}_{\mathsf{G}}^{L_p(w_\alpha)}(t) \geq \sup_{s>0} f_s^*(t) \geq c \sup_{s>t} s^{-\frac{1}{p_0}} \sim t^{-\frac{\alpha}{n} - \frac{1}{p}}, \quad t \to \infty. \qquad \square$$

Remark 10.18 Obviously, by Propositions 3.30, 3.35, the weight functions $w_\alpha(x) = \langle x \rangle^\alpha$, and $w^\alpha(x) = |x|^\alpha \in \mathcal{A}_p$ show the same influence on the underlying L_p-space as regards global assertions (integrability) in contrast to local assertions. It was certainly interesting to study this subject in a more general context, whereas one should expect rather qualitative characterisations then.

Theorem 10.19 *Let $0 < q \leq \infty$, $0 < p \leq \infty$ (with $p < \infty$ for $A = F$), and $s > \sigma_p$. Then*

$$\mathcal{E}_{\mathsf{G}}^{A_{p,q}^s}(t) \sim t^{-\frac{1}{p}}, \quad t \to \infty. \tag{10.17}$$

P r o o f : We shall present the argument for $p < \infty$ only; when $A = B$, everything can be extended to $p = \infty$ without difficulties. We start with the estimate from above and assume first $1 \leq p < \infty$. Then by (7.7), (7.26), (7.30), (8.56), (8.57) we have

$$A^s_{p,q} \hookrightarrow L_p, \tag{10.18}$$

so that Proposition 3.4(iv) and (10.14) imply the estimate from above,

$$\mathcal{E}_G^{A^s_{p,q}}(t) \leq c\, t^{-\frac{1}{p}}, \quad t \to \infty. \tag{10.19}$$

Dealing with $0 < p < 1$, the difficulty in extending (10.18) results from questions of convergence, i.e., that $A^s_{p,q}$ consists of tempered distributions $f \in \mathcal{S}'$, whereas this is not the case for L_p with $0 < p < 1$; cf. [Tri92, Rem. 2.3.2/3]. However, for $s > \sigma_p = n\left(\frac{1}{p} - 1\right)$, $0 < p < 1$, we always have at least $A^s_{p,q} \hookrightarrow L_1$. Furthermore, we can even estimate L_p-norms for $f \in A^s_{p,q}$ due to the following characterisation: Let $s > \sigma_p$, and $\varrho = \varphi_1 = \varphi(2^{-1}\cdot) - \varphi$, where φ is given by (7.1). Then

$$\|f|B^s_{p,q}\| \sim \|f|L_p\| + \left(\int_0^\infty t^{-sq} \left\| \mathcal{F}^{-1}\left(\varrho\left(t\cdot\right)\mathcal{F}f\right)|L_p\right\|^q \frac{dt}{t}\right)^{1/q}, \tag{10.20}$$

and

$$\|f|F^s_{p,q}\| \sim \|f|L_p\| + \left\| \left(\int_0^\infty t^{-sq} \left| \mathcal{F}^{-1}\left(\varrho\left(t\cdot\right)\mathcal{F}f\right)(\cdot)\right|^q \frac{dt}{t}\right)^{1/q} \Big| L_p\right\|, \tag{10.21}$$

cf. [Tri92, Rem. 2.3.3]. This immediately finishes the proof of the upper estimate for $0 < p < 1$: let $t > 0$ be arbitrarily large, then for each $f \in A^s_{p,q}$ with $\|f|A^s_{p,q}\| \leq 1$, (10.20) and (10.21), respectively, imply $\|f|L_p\| \leq c$, such that (10.14) leads to $f^*(t) \leq c'\, t^{-\frac{1}{p}}$, i.e., we obtain (10.19) for all p, $0 < p < \infty$.

For the converse, we make use of the equivalent norms (10.20), (10.21) again and a homogeneity argument. Let $R > 0$; then straightforward calculation shows that

$$\|f(R\cdot)|L_p\| = R^{-n/p}\|f|L_p\|$$

and

$$\mathcal{F}^{-1}\left(\varrho\left(t\cdot\right)\mathcal{F}\left[f(R\cdot)\right]\right)(x) = \mathcal{F}^{-1}\left(\varrho\left(t\cdot\right)\left[R^{-n}\mathcal{F}f\left(R^{-1}\cdot\right)\right]\right)(x)$$

$$= \mathcal{F}^{-1}\left(\varrho\left(Rt\cdot\right)\mathcal{F}f\right)(Rx), \quad x \in \mathbb{R}^n,$$

hence

$$\left\| \mathcal{F}^{-1}\left(\varrho\left(t\cdot\right)\mathcal{F}\left[f(R\cdot)\right]\right)|L_p\right\| = R^{-n/p}\left\| \mathcal{F}^{-1}\left(\varrho\left(Rt\cdot\right)\mathcal{F}f\right)|L_p\right\|$$

We apply (10.20) to $f_R = f(R\cdot)$ and thus obtain

$$\left\| f(R\cdot)|B^s_{p,q} \right\| \leq c_1 R^{-\frac{n}{p}} \left\{ \|f|L_p\| + \left(\int_0^\infty t^{-sq} \left\| \mathcal{F}^{-1} \left(\varrho\,(Rt\cdot)\,\mathcal{F}f \right)|L_p \right\|^q \frac{dt}{t} \right)^{\frac{1}{q}} \right\}$$

$$= c_1 R^{-\frac{n}{p}} \left\{ \|f|L_p\| + R^s \left(\int_0^\infty \tau^{-sq} \left\| \mathcal{F}^{-1} \left(\varrho\,(\tau\cdot)\,\mathcal{F}f \right)|L_p \right\|^q \frac{d\tau}{\tau} \right)^{\frac{1}{q}} \right\}$$

$$\leq c_2 \max\left(R^{-\frac{n}{p}}, R^{s-\frac{n}{p}} \right) \|f|B^s_{p,q}\|$$

With obvious modifications one can show that (10.21) implies

$$\left\| f(R\cdot)|F^s_{p,q} \right\| \leq c \max\left(R^{-\frac{n}{p}}, R^{s-\frac{n}{p}} \right) \|f|F^s_{p,q}\|,$$

so that we obtain for $0 < R \leq 1$,

$$\left\| f(R\cdot)|A^s_{p,q} \right\| \leq c\, R^{-\frac{n}{p}} \|f|A^s_{p,q}\|. \tag{10.22}$$

Assume now $0 < R \leq 1$ and consider functions

$$\psi_R(x) = R^{\frac{n}{p}}\, \psi\,(Rx), \quad x \in \mathbb{R}^n, \tag{10.23}$$

where ψ is given by (3.36). Then by the above estimates,

$$\left\| \psi_R|A^s_{p,q} \right\| \leq c \left\| \psi|A^s_{p,q} \right\| \leq c', \quad 0 < R \leq 1,$$

and $\psi_R^*(t) \sim R^{\frac{n}{p}}$ for $t \sim R^{-n}$, $0 < R \leq 1$. Let $t > 1$ be large, then $R_0 \sim t^{-1/n} \in (0,1)$, and up to possible normalisations this yields

$$\mathcal{E}_{\mathsf{G}}^{A^s_{p,q}}(t) \geq \sup_{0 < R < 1} \psi_R^*(t) \geq \psi_{R_0}^*(t) \sim R_0^{n/p} \sim t^{-1/p},$$

and the proof is finished. $\qquad\qquad\qquad\qquad\qquad\qquad\qquad\qquad \Box$

Remark 10.20 Obviously the assumption $s > \sigma_p$ is essentially used in the above argument. It is not yet clear what will happen in the borderline case $s = \sigma_p$ in the global setting.

Dealing with spaces of generalised smoothness $A^{(s,\Psi)}_{p,q}$, $0 < p < \infty$, $0 < q \leq \infty$, $s > \sigma_p$, and Ψ slowly varying, we obtain as a direct consequence of Proposition 10.19,

$$\mathcal{E}_{\mathsf{G}}^{A^{(s,\Psi)}_{p,q}}(t) \sim t^{-\frac{1}{p}}, \quad t \to \infty,$$

in view of Remark 7.21, in particular, (7.62).

We finally turn to weighted Besov spaces again; recall our local results Propositions 8.8 and 8.10. For convenience we restrict ourselves to the case $p > 1$.

Proposition 10.21 *Let* $1 < p < \infty$, $0 < q \leq \infty$, $s > 0$, *and* $0 \leq \alpha < \frac{n}{p'}$. *Then*

$$\mathcal{E}_G^{A_{p,q}^s(w_\alpha)}(t) \sim \mathcal{E}_G^{A_{p,q}^s(w^\alpha)}(t) \sim t^{-\frac{\alpha}{n}-\frac{1}{p}}, \quad t \to \infty. \tag{10.24}$$

P r o o f : We deal first with w_α and start with the estimate from above and combine for that reason the (weighted) embedding results (8.20) and Proposition 10.17. In particular, due to (8.20), which is based upon (8.19), we have the counterpart of (10.18),

$$A_{p,q}^s(w_\alpha) \hookrightarrow L_p(w_\alpha), \tag{10.25}$$

such that Proposition 3.4(iv) and (10.16) imply for $t \to \infty$,

$$\mathcal{E}_G^{A_{p,q}^s(w_\alpha)}(t) \leq c\, \mathcal{E}_G^{L_p(w_\alpha)}(t) \leq c'\, t^{-\frac{1}{p}-\frac{\alpha}{n}}.$$

The counterpart of (10.25) for w^α is covered by [Bui82, Thms. 2.6, 2.8],

$$A_{p,q}^s(w^\alpha) \hookrightarrow L_p(w^\alpha), \tag{10.26}$$

and Propositions 3.4(iv) and 10.17 complete the upper estimate in (10.24).

We prove the converse estimate and adapt the argument from the proof of Proposition 10.19 suitably. Instead of (10.23) we consider now functions

$$\varrho_R(x) = R^{\frac{n}{p}+\alpha}\, \varrho\,(Rx), \quad x \in \mathbb{R}^n, \quad 0 < R < 1, \tag{10.27}$$

with

$$\varrho = \varphi_1 = \varphi(2^{-1}\cdot) - \varphi, \tag{10.28}$$

and φ is given by (7.1). Then, obviously, $\operatorname{supp} \varrho \subset \{x \in \mathbb{R}^n : 1 < |x| < 4\}$, and hence, for small $0 < R < 1$,

$$w_\alpha\left(R^{-1}x\right) \sim R^{-\alpha}, \quad x \in \operatorname{supp} \varrho. \tag{10.29}$$

Thus by the above argument (10.22) for $0 < R < 1$, and (8.19),

$$\left\|\varrho_R |A_{p,q}^s(w_\alpha)\right\| \leq c_1\, R^{\frac{n}{p}+\alpha} \left\|w_\alpha \varrho\,(R\cdot)\,|A_{p,q}^s\right\|$$

$$\leq c_2\, R^\alpha \left\|w_\alpha\left(R^{-1}\cdot\right)\varrho|A_{p,q}^s\right\| \leq c_3\, \left\|\varrho|A_{p,q}^s\right\| \leq c_4$$

where for the penultimate inequality we applied (10.29) and some characterisation of spaces $A_{p,q}^s$ via local means, see [Tri92, Sect. 2.4.6, 2.5.3]. For convenience, assume that there is a number $\eta \in (1,2)$ such that $\varphi(x) < \frac{1}{2}$ for $|x| > \eta$, and $\varphi(x) > \frac{1}{2}$ for $|x| < \eta$. Consequently,

$$\left|\left\{x \in \mathbb{R}^n : \varrho(x) > \frac{1}{2}\right\}\right| \geq |\{x \in \mathbb{R}^n : \eta < |x| < 2\eta\}| \geq |\omega_n|\eta^n \geq |\omega_n|,$$

i.e.,

$$\varrho^*(|\omega_n|) \geq \frac{1}{2}. \tag{10.30}$$

Then $\varrho_R^*(t) \sim R^{\frac{n}{p}+\alpha} \varrho^*(R^{-n}t) \geq c\, R^{\frac{n}{p}+\alpha}$ for $t \sim R^{-n}$. For $t > 1$ large, choose $R_0 \sim t^{-1/n} \in (0,1)$, and up to possible normalisation this yields

$$\mathcal{E}_{\mathsf{G}}^{A_{p,q}^s(w_\alpha)}(t) \geq \sup_{0 < R < 1} \varrho_R^*(t) \geq \varrho_{R_0}^*(t) \sim R_0^{\frac{n}{p}+\alpha} \sim t^{-\frac{1}{p}-\frac{\alpha}{n}}.$$

It remains to deal with the weight $w^\alpha(x) = |x|^\alpha$. We consider functions

$$g_j(x) = 2^{-j\left(\frac{n}{p}+\alpha\right)} \varrho\left(2^{-j}x\right), \quad j \in \mathbb{N},$$

with ϱ given by (10.28). As calculated above this implies $g_j^*(t) \geq 2^{-j\frac{n}{p}-j\alpha}$ for $t \sim 2^{jn}$, $j \in \mathbb{N}$, leading to

$$\mathcal{E}_{\mathsf{G}}^{A_{p,q}^s(w^\alpha)}(t) \geq \sup_{j \in \mathbb{N}} g_j^*(t) \geq c\, t^{-\frac{1}{p}-\frac{\alpha}{n}}, \quad t \to \infty.$$

All that is left to prove is

$$\left\|g_j \mid A_{p,q}^s(w^\alpha)\right\| \leq c, \quad j \in \mathbb{N}.$$

Let k be a compactly supported C^∞ function on $\mathbb{R}^n$ with

$$\sum_{m \in \mathbb{Z}^n} k(x - m) = 1, \quad x \in \mathbb{R}^n.$$

Then we have for all $x \in \mathbb{R}^n$,

$$g_j(x) = 2^{-j\left(\frac{n}{p}+\alpha\right)} \sum_{m \in \mathbb{Z}^n} k(x - m)\, \varrho\left(2^{-j}x\right) \tag{10.31}$$

$$\sim 2^{-j\left(\frac{n}{p}+\alpha\right)} \sum_{|m| \sim 2^j} k(x - m)\, \varrho\left(2^{-j}x\right), \quad j \in \mathbb{N}. \tag{10.32}$$

On the other hand, $a_{0m}(x) := k(x - m)\, \varrho\left(2^{-j}x\right)$ can be regarded as 1_K-atoms located near Q_{0m}, $m \in \mathbb{Z}^n$, in the sense of Definition 7.7(i), and thus (10.32) represents a special atomic representation of g_j. By the already mentioned (Muckenhoupt) weighted version of Theorem 7.8 in [HPxx], [Bow05], this yields

$$\left\| g_j | A_{p,q}^s(w^\alpha) \right\|^p \leq c_1 \, 2^{-j\left(\frac{n}{p}+\alpha\right)p} \left\| \sum_{|m|\sim 2^j} \chi_{0m} | L_p\left(w^\alpha\right) \right\|^p$$

$$\leq c_2 \, 2^{-j\left(\frac{n}{p}+\alpha\right)p} \int_{\mathbb{R}^n} \left(\sum_{|m|\sim 2^j} \chi_{0m}(x) \right)^p |x|^{\alpha p} \mathrm{d}x$$

$$\leq c_3 \, 2^{-j\left(\frac{n}{p}+\alpha\right)p} \sum_{|m|\sim 2^j} \int_{Q_{0m}} |x|^{\alpha p} \mathrm{d}x$$

$$\leq c_4 \, 2^{-j\left(\frac{n}{p}+\alpha\right)p} \sum_{|m|\sim 2^j} |m|^{\alpha p}$$

$$\leq c_5 \, 2^{-j\left(\frac{n}{p}+\alpha\right)p+j\alpha p} \sum_{|m|\sim 2^j} 1$$

$$\leq c_6 \, 2^{-jn+jn} \leq c_6,$$

where χ_{0m} stands for the characteristic function of Q_{0m}, $m \in \mathbb{Z}^n$. This completes the argument. $\qquad\square$

Remark 10.22 Parallel to Remarks 10.16 and 10.18 we review our above global characterisations in comparison with the local (sub-critical) ones obtained in Section 8.1 and observe that in unweighted as well as weighted cases the additional smoothness $s > \sigma_p$ – in contrast to the underlying L_p-space – influences the local singularity behaviour unlike the global one. Moreover, the different weights $w_\alpha = \langle x \rangle^\alpha$ and $w^\alpha = |x|^\alpha$ cause globally the same asymptotic behaviour for the growth envelope function, whereas this is different locally. Avoiding limiting or borderline situations it also turns out that globally there is no difference between B- and F-spaces; in particular, there is no dependence on the fine index q.

Chapter 11

Applications

We present a few applications of our envelope results. Though our original intention to introduce envelopes of function spaces was certainly of a different flavour, concentrating essentially on some better understanding of rather complicated function spaces, our studies admit quite a number of interesting results: rather direct there is a link to Hardy-type inequalities and limiting embeddings, presented first. With the help of some rather astonishing "lift arguments" for envelopes, as described in Section 11.2, we can finally tackle problems of compactness, that is, asymptotic estimates for entropy and approximation numbers of compact embeddings, with surprisingly sharp results. In this section we present some general results, but otherwise exemplify the possible applications with a few cases only.

11.1 Hardy inequalities and limiting embeddings

As a first application we can derive some Hardy-type inequalities. This follows immediately from our envelope results above together with the monotonicity (4.2), see Propositions 4.1, 3.4(v) and 5.3(iv): there are constants $\varepsilon > 0$, $c, c' > 0$ such that for all $f \in X \subset L_1^{\mathrm{loc}}$ and $g \in Y \hookrightarrow C$,

$$\sup_{0 < t < \varepsilon} \varkappa(t) \, \frac{f^*(t)}{\mathcal{E}_{\mathsf{G}}^X(t)} \leq c \, \|f|X\|,$$

and

$$\sup_{0 < t < \varepsilon} \varkappa(t) \, \frac{\omega(g,t)}{t \, \mathcal{E}_{\mathsf{C}}^Y(t)} \leq c \, \|g|Y\|,$$

respectively, if, and only if, $\varkappa$ is bounded.

Corollary 11.1 *Let $\varepsilon > 0$ be small, $\varkappa(t)$ be a positive monotonically decreasing function on $(0, \varepsilon]$, and $0 < v \leq \infty$.*

(i) *Assume that $X \nrightarrow L_\infty$ is a (quasi-) normed function space on $\mathbb{R}^n$, satisfying the assumptions of Lemma 3.8; let $\mathcal{E}_{\mathsf{G}}^X$ be the corresponding* *growth envelope function. Then*

$$\left(\int_0^\varepsilon \left[\varkappa(t) \, \frac{f^*(t)}{\mathcal{E}_\mathsf{G}^X(t)} \right]^v \mu_\mathsf{G}(dt) \right)^{1/v} \leq c \, \|f|X\|$$

for some $c > 0$ and all $f \in X$ if, and only if, $\varkappa$ is bounded and $u_\mathsf{G}^X \leq v \leq \infty$, with the modification

$$\sup_{t \in (0,\varepsilon)} \varkappa(t) \, \frac{f^*(t)}{\mathcal{E}_\mathsf{G}^X(t)} \leq c \, \|f|X\| \qquad (11.1)$$

if $v = \infty$. In particular, if $\varkappa$ is an arbitrary non-negative function on $(0, \varepsilon]$, then (11.1) holds if, and only if, $\varkappa$ is bounded.

(ii) *Let $X \hookrightarrow C$ be a function space on $\mathbb{R}^n$ with (5.8), $X \not\hookrightarrow \mathrm{Lip}^1$ and continuity envelope function $\mathcal{E}_\mathsf{C}^X$. Then*

$$\left(\int_0^\varepsilon \left[\varkappa(t) \, \frac{\omega(f,t)}{t \, \mathcal{E}_\mathsf{C}^X(t)} \right]^v \mu_\mathsf{C}(dt) \right)^{1/v} \leq c \, \|f|X\|$$

for some $c > 0$ and all $f \in X$ if, and only if, $\varkappa$ is bounded and $u_\mathsf{C}^X \leq v \leq \infty$, with the modification

$$\sup_{t \in (0,\varepsilon)} \frac{\varkappa(t)}{\mathcal{E}_\mathsf{C}^X(t)} \, \frac{\omega(f,t)}{t} \leq c \, \|f|X\| \qquad (11.2)$$

if $v = \infty$. In particular, if $\varkappa$ is an arbitrary non-negative function on $(0, \varepsilon]$, then (11.2) holds if, and only if, $\varkappa$ is bounded.

Obviously we can establish a lot of assertions similar to Corollary 11.2 by a combination of Corollary 11.1 and our envelope results presented in the last sections. We restrict ourselves to an example only, related to spaces of type $A_{p,q}^s(w^\alpha)$, recall Proposition 8.10.

Corollary 11.2 *Let $0 < q \leq \infty$, $s > 0$, $1 < r < \infty$ and p with $0 < p < \infty$ be such that $s - \frac{n}{p} = -\frac{n}{r}$. Let $0 < \alpha < \frac{n}{r'}$. Let $\varkappa(t)$ be a positive monotonically decreasing function on $(0, \varepsilon]$ and let $0 < v \leq \infty$.*

(i) *Then*

$$\left(\int_0^\varepsilon \left[\varkappa(t) \, t^{\frac{1}{r} + \frac{\alpha}{n}} \, f^*(t) \right]^v \frac{dt}{t} \right)^{1/v} \leq c \, \|f|B_{p,q}^s(w^\alpha)\|$$

for some $c > 0$ *and all* $f \in B_{p,q}^s(w^\alpha)$ *if, and only if,* $\varkappa$ *is bounded and* $q \leq v \leq \infty$, *with the modification*

$$\sup_{t \in (0,\varepsilon)} \varkappa(t) \, t^{\frac{1}{r} + \frac{\alpha}{n}} \, f^*(t) \leq c \left\| f | B_{p,q}^s(w^\alpha) \right\| \tag{11.3}$$

if $v = \infty$. *In particular, if* $\varkappa$ *is an arbitrary non-negative function on* $(0, \varepsilon]$, *then* (11.3) *holds if, and only if,* $\varkappa$ *is bounded.*

(ii) *Then*

$$\left(\int_0^\varepsilon \left[\varkappa(t) \, t^{\frac{1}{r} + \frac{\alpha}{n}} \, f^*(t) \right]^v \frac{dt}{t} \right)^{1/v} \leq c \left\| f | F_{p,q}^s(w^\alpha) \right\|$$

for some $c > 0$ *and all* $f \in F_{p,q}^s(w^\alpha)$ *if, and only if,* $\varkappa$ *is bounded and* $p \leq v \leq \infty$, *with the modification*

$$\sup_{t \in (0,\varepsilon)} \varkappa(t) \, t^{\frac{1}{r} + \frac{\alpha}{n}} \, f^*(t) \leq c \left\| f | F_{p,q}^s(w^\alpha) \right\| \tag{11.4}$$

if $v = \infty$. *In particular, if* $\varkappa$ *is an arbitrary non-negative function on* $(0, \varepsilon]$, *then* (11.4) *holds if, and only if,* $\varkappa$ *is bounded.*

Remark 11.3 In [HM04, Cor. 3.4] and [CH05, Cor. 4.1] we obtained similar results when dealing with (unweighted) spaces of generalised smoothness $A_{p,q}^{(s,\Psi)}$.

Another application concerns necessary conditions for (limiting) embedding assertions. Propositions 3.4(iv) and 5.3(iii) as well as Propositions 4.5 and 6.4, respectively, lead to necessary conditions for embeddings

$$X_1 \hookrightarrow X_2. \tag{11.5}$$

We introduce some notation: For $X_i \subset L_1^{\mathrm{loc}}$ or $X_i \hookrightarrow C$, $i = 1, 2$, respectively, let

$$\mathsf{q}_\mathsf{G}^{(X_1, X_2)}(t) := \frac{\mathcal{E}_\mathsf{G}^{X_1}(t)}{\mathcal{E}_\mathsf{G}^{X_2}(t)}, \quad \mathsf{q}_\mathsf{C}^{(X_1, X_2)}(t) := \frac{\mathcal{E}_\mathsf{C}^{X_1}(t)}{\mathcal{E}_\mathsf{C}^{X_2}(t)}, \quad 0 < t < \varepsilon. \tag{11.6}$$

We may assume that $\varepsilon > 0$ is chosen sufficiently small, say, $\varepsilon \leq \tau_0^\mathsf{G}(X_2)$, given by (3.3), and $\varepsilon \leq \tau_0^\mathsf{C}(X_2)$, according to (5.2), such that (11.6) makes sense. Thus (11.5) and Propositions 3.4(iv) and 5.3(iii) imply that

$$\sup_{0 < t < \varepsilon} \mathsf{q}_\mathsf{G}^{(X_1, X_2)}(t) \leq c < \infty, \quad 0 < \varepsilon \leq \tau_0^\mathsf{G}(X_2), \tag{11.7}$$

or

$$\sup_{0 < t < \varepsilon} \mathsf{q}_\mathsf{C}^{(X_1, X_2)}(t) \leq c' < \infty, \quad 0 < \varepsilon \leq \tau_0^\mathsf{C}(X_2). \tag{11.8}$$

respectively. Moreover, Propositions 4.5 and 6.4 lead to

$$u_{\mathsf{G}}^{X_1} \le u_{\mathsf{G}}^{X_2} \quad \text{if} \quad X_1 \hookrightarrow X_2 \quad \text{and} \quad \mathsf{q}_{\mathsf{G}}^{(X_1,X_2)}(t) \sim 1, \; 0 < t < \varepsilon, \quad (11.9)$$

and

$$u_{\mathsf{C}}^{X_1} \le u_{\mathsf{C}}^{X_2} \quad \text{if} \quad X_1 \hookrightarrow X_2 \quad \text{and} \quad \mathsf{q}_{\mathsf{C}}^{(X_1,X_2)}(t) \sim 1, \; 0 < t < \varepsilon. \quad (11.10)$$

We exemplify this for some well-known situations first.

Example 11.4 We consider the embedding

$$B_{p_1,q_1}^{s_1}(\mathbb{R}^n) \hookrightarrow B_{p_2,q_2}^{s_2}(\mathbb{R}^n), \quad (11.11)$$

where $0 < p_i, q_i \le \infty$, $s_i \in \mathbb{R}$, $i = 1, 2$. Let $s_1 \ge s_2$; for convenience we assume $p_i < \infty$. Then we can easily derive the counterpart of (7.32), namely that (11.11) implies $p_1 \le p_2$, $\delta \ge 0$, and $q_1 \le q_2$ when $\delta = 0$, where we used notation (7.34). This can be seen as follows: Note first that in view of the lift result (7.8) we have (11.11) if, and only if,

$$B_{p_1,q_1}^{s_1-\sigma}(\mathbb{R}^n) \hookrightarrow B_{p_2,q_2}^{s_2-\sigma}(\mathbb{R}^n), \quad \sigma \in \mathbb{R}. \quad (11.12)$$

Choose first $\sigma < \min(s_1 - \sigma_{p_1}, s_2 - \sigma_{p_2})$, then we can apply Theorem 10.19 to $B_{p_i,q_i}^{s_i-\sigma}$, $i = 1, 2$, which yields $\tau_0^{\mathsf{G}}(B_{p_2,q_2}^{s_2-\sigma}) = \infty$, and together with (11.11), (11.12),

$$\mathsf{q}_{\mathsf{G}}^{(B_1^\sigma,B_2^\sigma)}(t) := \mathsf{q}_{\mathsf{G}}^{(B_{p_1,q_1}^{s_1-\sigma},B_{p_2,q_2}^{s_2-\sigma})}(t) \sim t^{\frac{1}{p_2}-\frac{1}{p_1}}, \quad t \to \infty. \quad (11.13)$$

Thus (11.7) leads to

$$\frac{1}{p_2} - \frac{1}{p_1} \le 0, \quad \text{i.e.,} \quad p_1 \le p_2. \quad (11.14)$$

Next we assume that

$$s_1 - \sigma_{p_1} > s_2 - \frac{n}{p_2} \quad (11.15)$$

and choose $\sigma \in \mathbb{R}$ in (11.12) such that

$$s_2 - \frac{n}{p_2} < \sigma < \min(s_1 - \sigma_{p_1}, s_2 - \sigma_{p_2}),$$

hence

$$s_1 - \sigma > \sigma_{p_1}, \quad \sigma_{p_2} < s_2 - \sigma < \frac{n}{p_2}. \quad (11.16)$$

Now we are prepared to complete the argument in this case, as by assumption (11.16) we can apply Theorems 8.1, 8.16(ii), and (7.41), (7.26) and Proposition 3.4(iii) to obtain for all small t, $0 < t < \varepsilon$,

$$\mathsf{q}_{\mathsf{G}}^{(B_1^\sigma,B_2^\sigma)}(t) \sim t^{-\frac{s_2-\sigma}{n}+\frac{1}{p_2}} \begin{cases} t^{\frac{s_1-\sigma}{n}-\frac{1}{p_1}}, & s_1 - \sigma < \frac{n}{p_1}, \; 0 < q_1 \le \infty \\ |\log t|^{\frac{1}{q_1}}, & s_1 - \sigma = \frac{n}{p_1}, \; 1 < q_1 \le \infty \\ c, & \text{otherwise} \end{cases}. \quad (11.17)$$

Thus (11.7) and (11.17) imply for all cases covered by (11.15),

$$s_1 - \sigma - \frac{n}{p_1} \geq s_2 - \sigma - \frac{n}{p_2},$$

i.e.,

$$\delta \geq 0. \tag{11.18}$$

In case of $\delta = 0$ we would like to conclude $q_1 \leq q_2$. For that reason note that our choice of σ as in (11.16) leads to $\mathfrak{q}_G^{(B_1^\sigma, B_2^\sigma)}(t) \sim 1$ in view of (11.17), such that (11.9) and Theorem 8.1 provide the additional condition

$$q_1 = u_G^{B_1^\sigma} \leq u_G^{B_2^\sigma} = q_2 \qquad \text{when} \quad \delta = 0, \tag{11.19}$$

i.e., the well-known (sufficient) conditions for the embedding (11.11). It remains to discuss the case not covered by (11.15), that is,

$$s_1 - \sigma_{p_1} \leq s_2 - \frac{n}{p_2}. \tag{11.20}$$

We disprove (11.11) in that case. Choose $\sigma \in \mathbb{R}$ in (11.12) such that

$$s_1 - \frac{n}{p_1} < \sigma < s_1 - \sigma_{p_1}, \tag{11.21}$$

then (11.20) immediately implies $s_2 - \sigma > \frac{n}{p_2} > \sigma_{p_2}$ and (7.41), (7.26), Proposition 3.4(iii), and Theorem 10.19 give

$$\mathcal{E}_G^{B_{p_2, q_2}^{s_2 - \sigma}}(t) \sim 1, \quad 0 < t < \varepsilon. \tag{11.22}$$

On the other hand, by (11.21) and Theorem 8.1 we conclude

$$\mathcal{E}_G^{B_{p_1, q_1}^{s_1 - \sigma}}(t) \sim t^{\frac{s_1 - \sigma}{n} - \frac{1}{p_1}}, \quad 0 < t < \varepsilon, \tag{11.23}$$

which led to a contradiction with (11.22) and Proposition 3.4(iv) if (11.12) and hence (11.11) was satisfied.

Remark 11.5 The cases $p_i = \infty$ could be incorporated using (11.8) for the local argument instead (with an appropriate lift, i.e., suitably chosen number σ), whereas the global one already covers this case. Moreover, instead of (11.9) one could argue in (11.19) by means of (11.7) and Theorem 8.16(ii), modifying the lift (11.16) properly (e.g., $s_2 = \frac{n}{p_2}$), at least for $q_i > 1$. In [CH05, Prop. 4.3] we used envelope results for spaces $A_{p,q}^{(s, \Psi)}$, see Remarks 8.18 and 9.7, to prove necessary conditions for the embedding

$$A_{p_1, q_1}^{(s_1, \Psi_1)} \hookrightarrow A_{p_2, q_2}^{(s_2, \Psi_2)}$$

with $s_1 \geq s_2$, $0 < p_1 \leq p_2 \leq \infty$, $0 < q_1, q_2 \leq \infty$, and Ψ_1, Ψ_2 admissible functions. In particular, we studied the limiting case $s_1 - \frac{n}{p_1} = s_2 - \frac{n}{p_2}$ only and obtained precisely the counterparts of the sufficiency results by Moura in [Mou02, Prop. 1.1.13(iv)-(vi)].

Example 11.6 We return to our Example 7.5 concerning spaces $B_{p,q}^{s,b}$. Let $s_1 \geq s_2$, $b_1, b_2 \in \mathbb{R}$, $0 < p_1 \leq p_2 \leq \infty$, and $0 < q_1, q_2 \leq \infty$. We assume that

$$s_1 - \frac{n}{p_1} = s_2 - \frac{n}{p_2}.$$

Our result [CH05, Prop. 4.3] reads in this case as

$$B_{p_1,q_1}^{s_1,b_1} \hookrightarrow B_{p_2,q_2}^{s_2,b_2} \quad \text{if, and only if,} \quad \left\{ \begin{array}{ll} b_1 - b_2 \geq 0 & , q_1 \leq q_2 \\ b_1 - b_2 > \frac{1}{q_2} - \frac{1}{q_1} & , q_1 > q_2 \end{array} \right\}. \quad (11.24)$$

This situation was already known, see [Leo98, Thm. 1]. Moreover, in the special setting $s_1 = s_2$, $p_1 = p_2$, let us denote $q := q_1$, $r := q_2$, $\alpha := -b_1$, $\beta := -b_2$, then (11.24) can be reformulated into

$$B_{p,q}^{s,-\alpha} \hookrightarrow B_{p,r}^{s,-\beta} \quad \text{if, and only if,} \quad \left\{ \begin{array}{ll} \beta \geq \alpha & , q \leq r \\ \beta - \frac{1}{r} > \alpha - \frac{1}{q} & , q > r \end{array} \right\},$$

see the parallel result Proposition 2.23 for Lipschitz spaces.

We return to weighted situations now and first review Lemma 3.33.

Corollary 11.7 Let $1 < p < \infty$, $0 \leq \alpha < \frac{n}{p'}$, and p_0 be given by $\frac{1}{p_0} = \frac{1}{p} + \frac{\alpha}{n}$. Let $0 < r, u \leq \infty$.

(i) Let $w_\alpha(x) = \langle x \rangle^\alpha$ be given by (3.48); then

$$L_p(w_\alpha) \hookrightarrow L_{r,u} \quad (11.25)$$

if, and only if,

$$p_0 \leq r \leq p, \quad \text{and} \quad \left\{ \begin{array}{l} p \leq u \leq \infty, r = p \quad \text{or} \quad r = p_0 \\ 0 < u \leq \infty, p_0 < r < p \end{array} \right\}. \quad (11.26)$$

(ii) Let $w^\alpha(x) = |x|^\alpha$ be given by (3.52); then

$$L_p(w^\alpha) \hookrightarrow L_{r,u} \quad (11.27)$$

if, and only if,

$$r = p_0, \quad \text{and} \quad u \geq p. \quad (11.28)$$

Proof: The sufficiency in (ii) is given by Lemma 3.33, in particular by (3.54), together with the monotonicity of Lorentz spaces $L_{r,u}$ in u. Concerning (i), we have

$$L_p(w_\alpha) \hookrightarrow L_p \quad \text{and} \quad L_p(w_\alpha) \hookrightarrow L_{p_0,p}, \quad (11.29)$$

by (3.50) and (3.54), respectively, which implies the sufficiency in (11.26) for $r = p$ and $r = p_0$, using again monotonicity of Lorentz spaces. Otherwise, in case of $p_0 < r < p$ we strengthen real interpolation arguments (3.55) for the target spaces in (11.29) to obtain

$$L_p(w_\alpha) \hookrightarrow (L_p, L_{p_0,p})_{\theta,u} = L_{r,u},$$

where $\theta \in (0,1)$ is chosen such that

$$\frac{1}{r} = \frac{1-\theta}{p} + \frac{\theta}{p_0},$$

and u is arbitrary. It remains to verify the necessity of (11.26) and (11.28); we use our envelope results. Assume (11.25); then Propositions 3.30 and 10.17, in particular, (3.49) and (10.16) give

$$\mathcal{E}_G^{L_p(w_\alpha)}(t) \sim \left\{ \begin{array}{ll} t^{-\frac{1}{p}} & , \; 0 < t < 1 \\ t^{-\frac{\alpha}{n}-\frac{1}{p}} & , \; t \to \infty \end{array} \right\}, \tag{11.30}$$

whereas we have for $L_{p_0,p}$ by Proposition 3.12, i.e., (3.22),

$$\mathcal{E}_G^{L_{r,u}}(t) \sim t^{-\frac{1}{r}}, \qquad t > 0. \tag{11.31}$$

Then (11.7), (11.25), (11.30) and (11.31) imply

$$\frac{1}{r} \geq \frac{1}{p}, \quad \text{and} \quad \frac{1}{r} \leq \frac{1}{p} + \frac{\alpha}{n} = \frac{1}{p_0},$$

i.e., $p_0 \leq r \leq p$. Moreover, when $p = r$, then (11.9), (11.30), (11.31) lead to

$$u_G^{L_p(w_\alpha)} = p \leq u = u_G^{L_{p,u}},$$

see Theorem 4.7(i) and Proposition 4.13. To complete the proof of (i) we have to show that

$$L_p(w_\alpha) \hookrightarrow L_{p_0,u} \tag{11.32}$$

implies $u \geq p$. We assume $u < p$ and disprove (11.32). For that purpose, consider the function

$$f_\gamma(x) = (1 + |\log|x||)^{-\gamma} \left\{ \begin{array}{ll} |x|^{-\frac{n}{p}} & , \; |x| < 1 \\ |x|^{-\frac{n}{p_0}} & , \; |x| > 1 \end{array} \right\}, \qquad x \in \mathbb{R}^n;$$

then easy calculations show that $f_\gamma \in L_p(w_\alpha)$ if, and only if, $\gamma > \frac{1}{p}$. On the other hand,

$$f_\gamma^*(t) \sim (1 + |\log t|)^{-\gamma} \left\{ \begin{array}{ll} t^{-\frac{1}{p}} & , \; t < 1 \\ t^{-\frac{1}{p_0}} & , \; t > 1 \end{array} \right\},$$

such that

$$\|f_\gamma|L_{p_0,u}\| \geq \left(\int\limits_2^\infty \left[t^{\frac{1}{p_0}} f_\gamma^*(t)\right]^u \frac{dt}{t}\right)^{1/u} \sim \left(\int\limits_2^\infty (\log t)^{-\gamma u} \frac{dt}{t}\right)^{1/u}$$

which does not converge for $\gamma \leq \frac{1}{u}$. Thus $f_\gamma \in L_p(w_\alpha) \setminus L_{p_0,u}$ for all γ with $\frac{1}{u} \leq \gamma < \frac{1}{p}$ in contrast to (11.32).

In case of (ii), we have to replace (11.30) by

$$\mathcal{E}_G^{L_p(w^\alpha)}(t) \sim t^{-\frac{\alpha}{n}-\frac{1}{p}}, \quad t > 0,$$

see Proposition 3.35, in particular (3.58). Following the above scheme, this immediately leads to $p_0 \leq r \leq p_0$ and $u \geq p$ when $r = p_0$, i.e., (11.28). □

We end this section with another consequence in the sense of the above Example 11.4, dealing now with weighted Besov spaces instead. We complete the characterisation (8.34), at least for $p_i > 1$.

Corollary 11.8 *Let* $1 < p_1 < \infty$, $1 < p_2 \leq \infty$, $0 < q \leq \infty$, $i = 1, 2$, $s_1 > s_2$, $0 < \alpha < \frac{n}{p_1'}$. *Then*

$$B_{p_1,q}^{s_1}(w^\alpha) \hookrightarrow B_{p_2,q}^{s_2} \tag{11.33}$$

implies

$$\alpha \leq \delta, \quad and \quad \frac{1}{p_2} \leq \frac{\alpha}{n} + \frac{1}{p_1}. \tag{11.34}$$

P r o o f: We proceed just as in the Example 11.4 above. Instead of the operator I_σ given by (7.6) we can use the operator J_σ introduced in [Bui82, Thm. 2.8],

$$J_\sigma = \mathcal{F}^{-1}\left(1 + 4\pi^2|\xi|^2\right)^{\sigma/2}, \quad \sigma \in \mathbb{R},$$

that maps

$$J_\sigma : A_{p,q}^s(\mathbb{R}^n, w) \longrightarrow A_{p,q}^{s-\sigma}(\mathbb{R}^n, w) \tag{11.35}$$

isomorphically for all weights $w \in \mathcal{A}_\infty$, $0 < p \leq \infty$, $0 < q \leq \infty$, $s \in \mathbb{R}$. This is the needed counterpart of (7.8). So we have (11.12) again; with the same choice of σ as above (now with $\sigma_{p_1} = \sigma_{p_2} = 0$), i.e., $\sigma < s_2$, we can apply Theorem 10.19 to $B_{p_2,q_2}^{s_2-\sigma}$, and Proposition 10.21 to $B_{p_1,q_1}^{s_1-\sigma}(w^\alpha)$, leading to $\tau_0^G(B_{p_2,q_2}^{s_2-\sigma}) = \infty$ again and substitute (11.13) by

$$\mathfrak{q}_G^{(B_{p_1,q_1}^{s_1-\sigma}(w^\alpha), B_{p_2,q_2}^{s_2-\sigma})}(t) \sim t^{\frac{1}{p_2}-\frac{1}{p_1}-\frac{\alpha}{n}}, \quad t \to \infty, \tag{11.36}$$

thus

$$\frac{1}{p_2} \leq \frac{1}{p_1} + \frac{\alpha}{n} \tag{11.37}$$

in view of (11.7). We study the local behaviour of $\mathsf{q}_{\mathsf{G}}^{(B_1^\sigma(w^\alpha),B_2^\sigma)}(t)$, using Theorem 8.1 and Proposition 8.10, more precisely, its extension discussed in Remark 8.11. We distinguish the following cases:

(i) $s_1 - \dfrac{n}{p_1} < s_2$ (ii) $s_2 \le s_1 - \dfrac{n}{p_1} < s_2 + \alpha$ (iii) $s_1 - \dfrac{n}{p_1} \ge s_2 + \alpha$

Starting with (i), choose σ in (11.35) such that

$$\max\left(s_1 - \frac{n}{p_1}, s_2 - \frac{n}{p_2}\right) < \sigma < s_2,$$

hence Proposition 8.10 and Theorem 8.1 give for small t, $0 < t < \varepsilon$,

$$\mathcal{E}_{\mathsf{G}}^{B_{p_1,q_1}^{s_1-\sigma}(w^\alpha)}(t) \sim t^{\frac{s_1-\sigma}{n} - \frac{1}{p_1} - \frac{\alpha}{n}}, \quad \mathcal{E}_{\mathsf{G}}^{B_{p_2,q_2}^{s_2-\sigma}}(t) \sim t^{\frac{s_2-\sigma}{n} - \frac{1}{p_2}}, \qquad (11.38)$$

respectively, that is,

$$\mathsf{q}_{\mathsf{G}}^{(B_1^\sigma(w^\alpha),B_2^\sigma)}(t) \sim t^{-\frac{\delta-\alpha}{n}}, \quad 0 < t < \varepsilon.$$

Thus (11.7) implies $\delta \ge \alpha$.

In case (ii) we can choose σ such that

$$\max\left(s_1 - \frac{n}{p_1} - \alpha, s_2 - \frac{n}{p_2}\right) < \sigma < \min\left(s_2, s_1 - \frac{n}{p_1} - \alpha + n\right).$$

Using the extended version of Proposition 8.10, i.e., (8.44) instead, we can argue in the same way as above and obtain (11.34).

Finally, we come to (iii); this leads directly to $\delta \ge s_2 + \alpha - s_2 + \dfrac{n}{p_2} \ge \alpha$. □

Remark 11.9 Comparison of (8.34) and (11.34) shows that by the envelope methods used above we only obtain $\frac{1}{p_2} \le \frac{1}{p_1} + \frac{\alpha}{n}$, whereas the necessary and sufficient condition reads as $\frac{1}{p_2} < \frac{1}{p_1} + \frac{\alpha}{n}$, see for instance [HS06] and [KLSS06b]. Moreover, there are extensions to different q-parameters, too, in the sense that (8.34) holds if, and only if, $\alpha \le \delta$ when $q_1 \le q_2$, and $\alpha < \delta$ for $q_1 > q_2$.

11.2 Envelopes and lifts

Recall that $\mathcal{E}_{\mathsf{G}}^X(t)$ is bounded when $X \hookrightarrow L_\infty$, see Proposition 3.4(iii), whereas $\mathcal{E}_{\mathsf{C}}^X(t)$ is only defined for $X \hookrightarrow C$. Thus it might not appear very interesting at first glance to study the interplay of $\mathcal{E}_{\mathsf{G}}^{X_1}$ and $\mathcal{E}_{\mathsf{C}}^{X_2}$ in general – at least not when the spaces X_1 and X_2 coincide, $X_1 = X_2$. We may, however, observe some phenomena granted that X_1 and X_2 are connected in a suitable way; we shall try to interpret and generalise this afterwards.

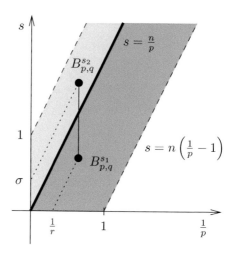

Figure 15

We consider the following situation. Let $0 < p < \infty$ and $0 < q \leq \infty$. Assume (as indicated in Figure 15) that $s_1 = \frac{n}{p} - \frac{n}{r}$ for some r, $1 < r < \infty$, and $s_2 = \sigma + \frac{n}{p}$ for some σ with $0 < \sigma < 1$. We consider the case that $s_2 = s_1 + 1$; that is, where $\sigma = 1 - \frac{n}{r}$. (Note that the assumptions on σ thus imply $r > n$.) Furthermore, by Theorem 8.1 we know $\mathcal{E}_{\mathsf{G}}^{A_{p,q}^{s}}(t) \sim t^{-\frac{1}{r}}$, whereas Theorem 9.2 yields $\mathcal{E}_{\mathsf{C}}^{A_{p,q}^{s+1}}(t) \sim t^{-(1-\sigma)}$. Consequently we obtain in that case

$$\mathcal{E}_{\mathsf{C}}^{A_{p,q}^{s+1}}(t) \sim t^{-(1-\sigma)} = (t^n)^{-\frac{1}{r}} \sim \mathcal{E}_{\mathsf{G}}^{A_{p,q}^{s}}(t^n).$$

Likewise, for $0 < p < n$ and $0 < q \leq 1$ Theorems 9.10 and 8.1 (with $r = n$) lead to

$$\mathcal{E}_{\mathsf{C}}^{A_{p,q}^{n/p}}(t) \sim t^{-1} = (t^n)^{-\frac{1}{r}} \sim \mathcal{E}_{\mathsf{G}}^{A_{p,q}^{n/p-1}}(t^n).$$

A similar behaviour can be observed when dealing with the borderline cases, $B_{p,q}^{n/p}$ and $B_{p,q}^{1+n/p}$, respectively,

$$\mathcal{E}_{\mathsf{C}}^{B_{p,q}^{1+n/p}}(t) \sim |\log t|^{\frac{1}{q'}} \sim \mathcal{E}_{\mathsf{G}}^{B_{p,q}^{n/p}}(t^n),$$

and a parallel result for the F-case. However, the log-function spoils the interplay of t and t^n in that case. Turning to the envelopes $\mathfrak{E}_G$ or $\mathfrak{E}_C$, it thus appears reasonable to define

$$\mathfrak{E}_G^n(X) := \left(\mathcal{E}_G^X(t^n), \, u_G^X \right),$$

where u_G^X is given as in Definition 4.2. Then Theorems 8.1 and 9.2, as well as Theorems 8.16 and 9.4 lead to

$$\mathfrak{E}_G^n(A_{p,q}^s) = \mathfrak{E}_C\left(A_{p,q}^{s+1}\right) \quad \text{if} \quad \left\{ \begin{array}{l} 0 < p < \infty, \ s = \frac{n}{p} - \frac{n}{r}, \ n < r < \infty \\ 1 < p < \infty, \ s = \frac{n}{p}, \qquad A_{p,q}^s = F_{p,q}^s \\ 1 < q \le \infty, \ s = \frac{n}{p}, \qquad A_{p,q}^s = B_{p,q}^s \end{array} \right\}, \quad (11.39)$$

where we assume in general $0 < p,q \le \infty$. When $r = n$, i.e., $s = \frac{n}{p} - 1$, we have at least the corresponding result for the envelope functions,

$$\mathcal{E}_G^{A_{p,q}^s}(t^n) \sim \mathcal{E}_C^{A_{p,q}^{s+1}}(t), \qquad (11.40)$$

see Theorems 8.1 (with $r = n$) and 9.10.

Does this reflect a more general behaviour, that is, in what sense can this particular result be extended?

So far we only collected results "associated" in the above sense, but achieved (almost) independently of each other. The more desirable was a direct link between $\frac{\omega(f,t)}{t}$ and $|\nabla f|^*(t^n)$ or $|\nabla f|^{**}(t^n)$) for, say, $f \in X \hookrightarrow C^1$. We return to Proposition 5.10, in particular to estimate (5.11),

$$\omega(f,t) \le c \int_0^{t^n} s^{\frac{1}{n}-1} |\nabla f|^*(s)\mathrm{d}s \qquad (11.41)$$

for $t > 0$ and all $f \in C^1(\mathbb{R}^n)$. Plainly, this estimate plays an essential role in our subsequent study of $\mathcal{E}_C^{X_1}$ and $\mathcal{E}_G^{X_2}$, where $X_1 \hookrightarrow C$ and $X_2 \subset L_1^{\mathrm{loc}}$ are such that $|\nabla f| \in X_2$ for $f \in X_1$ (this setting is motivated by our above observations). We first discuss the "optimality" of (11.41). Recall that we have by (11.41) for $n = 1$,

$$\frac{\omega(f,t)}{t} \le c \, |f'|^{**}(t), \quad 0 < t < \varepsilon, \quad f \in C^1(\mathbb{R}). \qquad (11.42)$$

So one can ask whether a replacement of (5.11) in the sense of (11.42), i.e.,

$$\frac{\omega(f,t)}{t} \le c \, |\nabla f|^{**}(t^n), \quad 0 < t < \varepsilon, \qquad (11.43)$$

was true for all $f \in C^1(\mathbb{R}^n)$ and dimension $n > 1$. Obviously, (11.43) was sharper than (11.41), and also implied Triebel's result [Tri01, Prop. 12.16] mentioned in Remark 5.11,

$$\frac{\omega(f,t)}{t} \le c \, |\nabla f|^{**} \left(t^{2n-1}\right) + 3 \sup_{0<\tau\le t^2} \tau^{-\frac{1}{2}} \, \omega(f,\tau) \qquad (11.44)$$

for some small $\varepsilon > 0$ and all $0 < t < \varepsilon$ and all $f \in C^1(\mathbb{R}^n)$; we refer to [Har01, Sect. 6.3]. However, (11.43) cannot hold in general when $n > 1$; we give some argument disproving (11.43).

Assume (11.43) was true for $n > 1$. Let $f \in W_n^1(\mathbb{R}^n) = F_{n,2}^1(\mathbb{R}^n)$; by density arguments we may furthermore suppose that $f \in F_{n,2}^1(\mathbb{R}^n) \cap C_0^\infty(\mathbb{R}^n)$. Then by (7.9) $|\nabla f| \in F_{n,2}^0 = L_n$, leading to $|\nabla f|^{**}(\tau) \le C_n \, \tau^{-\frac{1}{n}}$, $\tau > 0$, and (11.43) then implies

$$\omega(f,t) \le c \, t \, |\nabla f|^{**}(t^n) \le c't \, (t^n)^{-\frac{1}{n}} = c'$$

for small $t > 0$. In other words, all $f \in F_{n,2}^1(\mathbb{R}^n) \cap C_0^\infty(\mathbb{R}^n)$ (and by the usual density arguments then all $f \in F_{n,2}^1(\mathbb{R}^n)$, too) are (locally) bounded. This, however, is wrong: recall (7.40) with $p = n > 1$; cf. Proposition 7.13. On the other hand, one can also rely on a result of Stein in [Ste81] stating that if a function f on $\mathbb{R}^n$ satisfies $\nabla f \in L_{n,1}$ locally, then f is equi-measurable with a continuous function, see also [DS84]. Moreover, there is a remark that the result is sharp in the following sense: taking $g \notin L_{n,1}$ with $f = |x|^{-(n-1)} * g$, then there is a positive $\tilde{g}$, equi-measurable with $|g|$, such that the resulting f is unbounded near every point; see also [Ste70, Ch. 8] and [Kol89, §5] for further details. So (11.41) – stating exactly that $|\nabla f|$ belongs to $L_{n,1}$ locally – is the best possible result (in that sense) and (11.43) – referring to $|\nabla f| \in L_n$ – cannot hold. The essential difference to the one-dimensional case is obvious in this setting as $L_{1,1} = L_1$, but $L_{p,1}(\mathbb{R}^n)$ is properly contained in $L_p(\mathbb{R}^n)$ for any $p > 1$.

Hence for $n > 1$ we are left with the two estimates (11.41) and (11.44) (instead of (11.43)) and try to compare them. At first glance it seems that our estimate (11.41) might be slightly sharper: though both estimates in question gave rise to the estimate (5.12), only (11.41) implies (5.13). The case $n = 1$ is clear: the second term in (11.44) disappears and we have (11.42) again.

Lemma 11.10 *Let $n > 1$. There is a $c > 0$ such that for all $0 < t < 1$ and all $f \in C^1(\mathbb{R}^n)$,*

$$\int_0^{t^n} s^{\frac{1}{n}-1} \, |\nabla f|^*(s)\mathrm{d}s \le c_1 \, t \, |\nabla f|^{**}\left(t^{2n-1}\right) + c_2 \, t^{2-\frac{1}{n}} \, \||f|C^1\|. \qquad (11.45)$$

P r o o f: We split the integral on the left-hand side of (11.45) as follows,

$$\int_0^{t^n} s^{\frac{1}{n}-1} |\nabla f|^*(s)\mathrm{d}s = \int_0^{t^{2n-1}} s^{\frac{1}{n}-1} |\nabla f|^*(s)\mathrm{d}s + \int_{t^{2n-1}}^{t^n} s^{\frac{1}{n}-1} |\nabla f|^*(s)\mathrm{d}s. \qquad (11.46)$$

We deal with the second term first and use the monotonicity of the maximal function g^{**} as well as $g^*(s) \le g^{**}(s)$,

$$\int_{t^{2n-1}}^{t^n} s^{\frac{1}{n}-1} |\nabla f|^* (s) ds \le |\nabla f|^{**} \left(t^{2n-1}\right) \int_{t^{2n-1}}^{t^n} s^{\frac{1}{n}-1} ds \le c\, t\, |\nabla f|^{**} \left(t^{2n-1}\right).$$

It remains to consider the first term on the right-hand side of (11.46); we verify that

$$\int_0^{t^{2n-1}} s^{\frac{1}{n}-1} |\nabla f|^* (s) ds \le c\, t^{2-\frac{1}{n}} \left\| f|C^1 \right\|.$$

This is an immediate consequence of our assumption $f \in C^1$,

$$|\nabla f|^* (0) = \| |\nabla f| |L_\infty\| \le c \left\| f|C^1 \right\|.$$

Hence the monotonicity of $|\nabla f|^* (s)$ implies

$$\int_0^{t^{2n-1}} s^{\frac{1}{n}-1} |\nabla f|^* (s) ds \le c \left\| f|C^1 \right\| \int_0^{t^{2n-1}} s^{\frac{1}{n}-1} ds = c'\, t^{2-\frac{1}{n}} \left\| f|C^1 \right\|.$$

The lemma is proved. $\qquad\qquad\qquad\qquad\qquad\qquad\qquad\qquad\qquad\square$

Obviously the estimate for the second term on the right-hand side in (11.45) is very rough and can probably be improved. On the other hand, a second term is surely necessary in general; for assume we (only) had

$$\int_0^{t^{2n-1}} s^{\frac{1}{n}-1} |\nabla f|^* (s) ds \le c\, t\, |\nabla f|^{**} \left(t^{2n-1}\right),$$

for all small $t > 0$ and $f \in C^1$, i.e., (by the definition of the maximal function g^{**})

$$\int_0^{t^{2n-1}} s^{\frac{1}{n}-1} |\nabla f|^* (s) ds \le c\, t^{-(2n-2)} \int_0^{t^{2n-1}} |\nabla f|^* (s) ds. \qquad (11.47)$$

Now a simple example of a function f with $|\nabla f|^* (s) \sim s^{-\varkappa}$ disproves (11.47) if we choose $\frac{1}{n} < \varkappa < 1$: then the left-hand side of (11.47) diverges whereas the right-hand side does not. So a second term is needed for "compensation" in general. Of course, splitting the integral in (11.46) not with $t^{2n-1} < t^n$ as an intermediate point, but with, say, $t^{n+\varepsilon} < t^n$, one can improve the first term on the right-hand side of (11.45) at the expense of the latter one,

$$\int\limits_0^{t^n} s^{\frac{1}{n}-1} \, |\nabla f|^* (s) \mathrm{d}s \le c_\varepsilon \left\{ t \, |\nabla f|^{**} \left(t^{n+\varepsilon} \right) + t^{1+\varepsilon} \, \|f|C^1\| \right\} ;$$

this argument resembles [Tri01, Rem. 12.17].

Note, that (11.44) leads to a right-hand side like (11.45), but with $t^{2-\frac{1}{n}}$ in the latter term replaced by t^2, which is smaller for $0 < t < 1$. So a combination of (11.41) and (11.45) results in an estimate less sharp than (11.44), but due to the partly rather rough arguments is it not clear at the moment, whether (11.44) or (11.41) are better in general. Nevertheless, for our purpose estimate (11.41) was completely sufficient; recall Proposition 5.10.

We come back to our "lifting" problem for the envelopes. Let $X \subset L_1^{loc}$ be some function space on $\mathbb{R}^n$ of regular distributions with, say, $X \not\hookrightarrow L_\infty$. Denote by $X^\nabla \subset X$ the following subspace

$$X^\nabla = \left\{ g \in L_1^{loc} \, : \, \mathrm{D}^\alpha g \in X, \ |\alpha| \le 1 \right\} \tag{11.48}$$

with

$$\|g|X^\nabla\| \sim \|g|X\| + \sum_{|\alpha|=1} \|\mathrm{D}^\alpha g \, |X\| .$$

We assume that $X^\nabla \hookrightarrow C$; this setting is obviously motivated by $X = A_{p,q}^s$, see (9.23). In view of (11.39) and (11.40) we study the problem under which assumptions one has

$$\mathfrak{E}_\mathsf{G}^n (X) = \mathfrak{E}_\mathsf{C} \left(X^\nabla \right)$$

or, at least,

$$\mathcal{E}_\mathsf{G}^X (t^n) \sim \mathcal{E}_\mathsf{C}^{X^\nabla} (t), \quad 0 < t < \varepsilon.$$

We have no complete answer, but a partial one. Let $f \in X^\nabla \cap C^1$ be such that $\left\| \, |\nabla f| \, |X\| \right\| \le \|f|X^\nabla\| \sim 1$ and $t > 0$ small. Then by (11.41),

$$\frac{\omega(f,t)}{t} \le c \, \frac{1}{t} \int\limits_0^{t^n} s^{\frac{1}{n}-1} |\nabla f|^* (s) \mathrm{d}s \le c \, \frac{1}{t} \int\limits_0^{t^n} s^{\frac{1}{n}-1} \mathcal{E}_\mathsf{G}^X (s) \mathrm{d}s$$

for all $f \in X^\nabla \cap C^1$, $\|f|X^\nabla\| = 1$, and small $t > 0$. Assuming further that, for instance, C_0^∞ is dense in X^∇, then this implies

$$\mathcal{E}_\mathsf{C}^{X^\nabla} (t) \le c \, \frac{1}{t} \int\limits_0^{t^n} s^{\frac{1}{n}-1} \mathcal{E}_\mathsf{G}^X (s) \mathrm{d}s \sim \frac{1}{t} \int\limits_0^{t} \mathcal{E}_\mathsf{G}^X (\sigma^n) \, \mathrm{d}\sigma. \tag{11.49}$$

In view of our above-mentioned examples (11.39) we would like to estimate (11.49) further by $c\,\mathcal{E}_\mathsf{G}^X\,(t^n)$; this refers to the question whether

$$\frac{1}{t}\int_0^t \mathcal{E}_\mathsf{G}^X\,(\sigma^n)\;\mathrm{d}\sigma \;\leq\; c\,\mathcal{E}_\mathsf{G}^X\,(t^n) \tag{11.50}$$

is true for some $c > 0$, and all small $t > 0$. Reformulating (11.50) we thus achieved the convergence of

$$\sum_{k=0}^\infty 2^{-k}\,\frac{\mathcal{E}_\mathsf{G}^X\left(2^{-(k+J)n}\right)}{\mathcal{E}_\mathsf{G}^X\left(2^{-Jn}\right)} \;\leq\; C \tag{11.51}$$

uniformly with respect to $J \in \mathbb{N}$, for, say, $J \geq J_0$, implies

$$\mathcal{E}_\mathsf{C}^{X^\nabla}(t) \;\leq\; c\,\mathcal{E}_\mathsf{G}^X\,(t^n) \tag{11.52}$$

for all small t, $0 < t < \varepsilon$. Clearly (11.51) is satisfied for

$$\mathcal{E}_\mathsf{G}^X(\tau) \;\sim\; \tau^{-\varkappa}\,|\log\tau|^\mu \qquad \text{with} \qquad \begin{cases} 0 < \varkappa < \frac{1}{n}\,, \mu \in \mathbb{R}, \\ \varkappa = 0 \quad\;, \mu > 0, \\ \varkappa = \frac{1}{n} \quad, \mu < -1\,; \end{cases} \tag{11.53}$$

this covers all cases in (11.39) apart from the limiting case when $X = B_{p,q}^{n/p-1}$, $X^\nabla = B_{p,q}^{n/p}$, $0 < p < n$, $0 < q \leq 1$, – reflecting that (11.51) is only sufficient for (11.52).

Turning to the indices u_G^X and $u_\mathsf{C}^{X^\nabla}$, respectively, one concludes

$$u_\mathsf{C}^{X^\nabla} \;\leq\; u_\mathsf{G}^X \tag{11.54}$$

provided that there is some $c > 0$ such that for all $k \in \mathbb{N}$,

$$\sum_{\nu=0}^{k-J} 2^{-\nu\varrho}\left[\frac{\mathcal{E}_\mathsf{G}^X\left(2^{-kn}\right)}{\mathcal{E}_\mathsf{C}^{X^\nabla}\left(2^{-(k-\nu)}\right)}\right]^r \;\leq\; c, \tag{11.55}$$

where $r = u_\mathsf{G}^X$ and $\varrho < r$ (in case of $r = u_\mathsf{G}^X \leq 1$ we may admit $\varrho = r$). The proof of this fact copies that one of Proposition 5.10, Step 3. Note that (5.12) and (5.13) are certain examples: the first one with $\mathcal{E}_\mathsf{C}^{X^\nabla}(t) \sim \mathcal{E}_\mathsf{G}^X(t) \sim |\log t|^u$ is performed directly in Step 3 of the proof of Proposition 5.10, whereas (5.13) is related to the setting $\mathcal{E}_\mathsf{C}^{X^\nabla}(t) \sim t^{-(1-\varkappa)}$, $\mathcal{E}_\mathsf{G}^X(t) \sim t^{-\frac{1}{n}(1-\varkappa)}$, $0 < \varkappa < 1$. Obviously condition (11.55) is satisfied in that case, too,

$$\sum_{\nu=0}^{k-J} 2^{-\nu\varrho}\left[\frac{\mathcal{E}_\mathsf{G}^X\left(2^{-kn}\right)}{\mathcal{E}_\mathsf{C}^{X^\nabla}\left(2^{-(k-\nu)}\right)}\right]^r \;\sim\; \sum_{\nu=0}^{k-J} 2^{-\nu\varrho}\left[\frac{2^{k(1-\varkappa)}}{2^{(k-\nu)(1-\varkappa)}}\right]^r$$

$$\sim\; \sum_{\nu=0}^{k-J} 2^{-\nu(\varrho-r(1-\varkappa))} \;\leq\; c$$

if we choose ϱ such that $r(1-\varkappa)<\varrho<r$.

Assume we knew already that $\mathcal{E}_{\mathsf{C}}^{X^{\nabla}}(t) \sim \mathcal{E}_{\mathsf{G}}^{X}(t^n)$ for small $t>0$. Then condition (11.55) can be reformulated as

$$\sum_{\nu=0}^{k-J} 2^{-\nu\varrho}\left[\frac{\mathcal{E}_{\mathsf{C}}^{X^{\nabla}}\left(2^{-k}\right)}{\mathcal{E}_{\mathsf{C}}^{X^{\nabla}}\left(2^{-(k-\nu)}\right)}\right]^r \le c \tag{11.56}$$

for some $c>0$ independent of $k \in \mathbb{N}$. Recall our difficulties with the situation of continuity envelopes on the critical line in Section 9.3; this refers to the situation $X=B_{p,q}^s$, $X^{\nabla}=B_{p,q}^{s+1}$, with $n>1$, $0<p<n$, $s=\frac{n}{p}-1$, $0<q\le1$, and

$$\mathcal{E}_{\mathsf{G}}^{B_{p,q}^s}(t^n) \sim \mathcal{E}_{\mathsf{C}}^{B_{p,q}^{s+1}}(t) \sim t^{-1},$$

and a similar result for F-spaces; see Theorems 8.1 (with $r=n$) and 9.10. Consequently (11.56) reads as the question whether

$$\sum_{\nu=0}^{k} 2^{-\nu\varrho}\left[\frac{\mathcal{E}_{\mathsf{C}}^{X^{\nabla}}\left(2^{-k}\right)}{\mathcal{E}_{\mathsf{C}}^{X^{\nabla}}\left(2^{-(k-\nu)}\right)}\right]^r \sim \sum_{\nu=0}^{k} 2^{-\nu\varrho}\left[\frac{2^k}{2^{(k-\nu)}}\right]^r = \sum_{\nu=0}^{k} 2^{-\nu(\varrho-r)}$$

converges independently of $k \in \mathbb{N}$. This, however, fails because of $\varrho \le r$. So condition (11.56) reflects the additional problems appearing on the critical line exactly.

Nevertheless we shall finally collect the above considerations for further use, though the answer is not yet complete – missing links "converse" to (11.49) and (11.54) in general case so far.

Corollary 11.11 *Let the spaces* X, X^{∇} *be given as above.*

(i)   *There exists* $c>0$ *such that*

$$\mathcal{E}_{\mathsf{C}}^{X^{\nabla}}(t) \le c\,\frac{1}{t}\int_0^{t^n} s^{\frac{1}{n}-1}\,\mathcal{E}_{\mathsf{G}}^{X}(s)ds. \tag{11.57}$$

for all small t, $0<t<\varepsilon$. *Moreover, if there is a number* $C>0$ *such that for all large* $J\in\mathbb{N}$, $J\ge J_0$,

$$\sum_{k=0}^{\infty} 2^{-k}\frac{\mathcal{E}_{\mathsf{G}}^{X}\left(2^{-(k+J)n}\right)}{\mathcal{E}_{\mathsf{G}}^{X}\left(2^{-Jn}\right)} \le C, \tag{11.58}$$

then (11.57) *can be replaced by*

$$\mathcal{E}_{\mathsf{C}}^{X^{\nabla}}(t) \le c\,\mathcal{E}_{\mathsf{G}}^{X}(t^n). \tag{11.59}$$

(ii)   *Assume there is a number* $c > 0$ *such that for all* $k \in \mathbb{N}$,

$$\sum_{\nu=0}^{k} 2^{-\nu \varrho} \left[\frac{\mathcal{E}_{\mathsf{G}}^{X}\left(2^{-kn}\right)}{\mathcal{E}_{\mathsf{C}}^{X^{\nabla}}\left(2^{-(k-\nu)}\right)} \right]^{r} \leq c, \qquad (11.60)$$

where $r = u_{\mathsf{G}}^{X}$ *and* $\varrho < r$ *(in case of* $r = u_{\mathsf{G}}^{X} \leq 1$ *we may admit* $\varrho = r$*). Then*

$$u_{\mathsf{C}}^{X^{\nabla}} \leq u_{\mathsf{G}}^{X}. \qquad (11.61)$$

In particular, when $\mathcal{E}_{\mathsf{C}}^{X^{\nabla}}(t) \sim \mathcal{E}_{\mathsf{G}}^{X}(t^{n})$, *(11.60) can be replaced by*

$$\sum_{\nu=0}^{k} 2^{-\nu \varrho} \left[\frac{\mathcal{E}_{\mathsf{C}}^{X^{\nabla}}\left(2^{-k}\right)}{\mathcal{E}_{\mathsf{C}}^{X^{\nabla}}\left(2^{-(k-\nu)}\right)} \right]^{r} \leq c.$$

Inequalities converse to (11.52) and (11.54) are missing so far; further studies in the sense of [JMP91] are necessary, and – in view of our results (11.39), (11.40) – also promising.

Remark 11.12 Note that (11.58) and (11.60) are only sufficient to get (11.59) and (11.61), respectively, but not necessary; recall the situation for $X = B_{p,q}^{n/p}$, $0 < p < \infty$, $1 < q \leq \infty$, or $X = W_{1}^{n}$, $X = B_{p,q}^{n/p-1}$, $0 < p < n$, $0 < q \leq 1$, for instance.

11.3   Compact embeddings

We briefly discuss some questions related to compactness. We already mentioned that – turning to spaces on bounded domains defined by restriction – most of our results for (growth or continuity) envelopes can be transferred immediately. For convenience, we shall only regard the unit ball $U \subset \mathbb{R}^{n}$ as an underlying domain in this section. We study an embedding between two function spaces defined on U, and possible links between its compactness and the envelopes of the involved spaces. Recall our notation (11.6). Clearly, by Propositions 3.4(iv) and 5.3(iii) there cannot be a continuous embedding $X_{1} \hookrightarrow X_{2}$ at all whenever

$$\sup_{0<t<\varepsilon} \mathsf{q}_{\mathsf{G}}^{(X_{1},X_{2})}(t) = \infty, \quad \text{or} \quad \sup_{0<t<\varepsilon} \mathsf{q}_{\mathsf{C}}^{(X_{1},X_{2})}(t) = \infty. \qquad (11.62)$$

So for a continuous embedding (not to speak of compactness so far) it is at least necessary that $\mathsf{q}_{\mathsf{G}}^{(X_{1},X_{2})}(t)$ or $\mathsf{q}_{\mathsf{C}}^{(X_{1},X_{2})}(t)$ are bounded, see also the discussion on "limiting" embeddings in Section 11.1. Moreover, granted the embedding $X_{1} \hookrightarrow X_{2}$ was continuous, the boundedness of $\mathsf{q}_{\mathsf{G}}^{(X_{1},X_{2})}(t)$, $\mathsf{q}_{\mathsf{C}}^{(X_{1},X_{2})}(t)$

is not sufficient for its compactness: Triebel claimed in [Tri01, 14.6] that, roughly speaking, some embedding cannot be compact when the envelopes of source and target spaces coincide, i.e., $q_G^{(X_1,X_2)}(t) \sim 1$ or $q_C^{(X_1,X_2)}(t) \sim 1$. He gave a proof for a special example; we shall generalise this now.

Proposition 11.13 *Let $X_1(U)$, $X_2(U)$ be function spaces of regular distributions with $X_1(U) \hookrightarrow X_2(U)$.*

(i)   *Assume that $C^\infty(\overline{U})$ is dense in $X_2(U)$, $X_2(U) \hookrightarrow C(U)$, $X_2(U) \not\hookrightarrow \mathrm{Lip}^1(U)$, and $q_C^{(X_1,X_2)}(t) \sim 1$, $0 < t < \varepsilon$. Then id $: X_1(U) \longrightarrow X_2(U)$ cannot be compact.*

(ii) *Assume that there is some $c > 0$ such that for small $t > 0$ we have $\mathcal{E}_G^{X_2}(t) \leq c\, \mathcal{E}_G^{X_2}(2t)$, let the simple functions be dense in $X_2(U)$, $X_2(U) \not\hookrightarrow L_\infty(U)$, and $q_G^{(X_1,X_2)}(t) \sim 1$, $0 < t < \varepsilon$. Then id $: X_1(U) \longrightarrow X_2(U)$ cannot be compact.*

P r o o f: Let $q_C^{(X_1,X_2)}(t) \sim 1$, i.e., $\mathcal{E}_C^{X_1}(t) \sim \mathcal{E}_C^{X_2}(t)$, $0 < t < \varepsilon$, and assume id $: X_1(U) \longrightarrow X_2(U)$ was compact; then for all $\gamma > 0$ we have a finite γ-net $\{f_1,\dots,f_{N(\gamma)}\} \subset X_2$, $\|f_j|X_2\| \leq 1$, such that for all $f \in X_1$, $\|f|X_1\| \leq 1$, it holds

$$\min_{j=1,\dots,N(\gamma)} \|f - f_j|X_2\| < \gamma,$$

where we may assume additionally that $f_j \in C^\infty(\overline{U})$, $j = 1,\dots,N(\gamma)$. Then by Proposition 5.3(iv), or (6.2) and the monotonicity of the left-hand sides of this equation in v, respectively, we obtain for small $\delta > 0$,

$$\sup_{0<t<\delta} \frac{\omega(f - f_j, t)}{t\, \mathcal{E}_C^{X_2}(t)} \leq c \|f - f_j|X_2\| < \gamma, \tag{11.63}$$

i.e., uniformly for all t, $0 < t < \delta$,

$$\frac{\omega(f,t)}{t} < \gamma\, \mathcal{E}_C^{X_2}(t) + \frac{\omega(f_j,t)}{t} \leq \gamma\, \mathcal{E}_C^{X_2}(t) + c_\delta, \tag{11.64}$$

because of the smoothness of f_j. Take the supremum over $f \in X_1$, $\|f|X_1\| \leq 1$, and restrict γ to small values, then we obtain

$$\mathcal{E}_C^{X_1}(t) \leq \gamma\, \mathcal{E}_C^{X_2}(t) + c_\delta, \quad 0 < t < \delta, \tag{11.65}$$

and thus by our assumption $\mathcal{E}_C^{X_1}(t) \sim \mathcal{E}_C^{X_2}(t)$,

$$\mathcal{E}_C^{X_2}(t) \leq c_{\delta,\gamma}, \quad 0 < t < \delta.$$

This, however, contradicts $X_2(U) \not\hookrightarrow \mathrm{Lip}^1(U)$ by Proposition 5.3(ii).

Dealing with the situation in (ii), (11.63) has to be replaced by

$$\sup_{0<t<\delta} \frac{(f-f_j)^*(t)}{\mathcal{E}_{\mathsf{G}}^{X_2}(t)} \leq c\,\|f-f_j|X_2\| < \gamma,$$

based on Proposition 3.4(v), or (4.5), respectively. Hence by the special sub-additivity of the non-increasing rearrangement and our additional assumption we arrive at

$$f^*(t) < \gamma\,\mathcal{E}_{\mathsf{G}}^{X_2}\left(\frac{t}{2}\right) + f_j^*\left(\frac{t}{2}\right) \leq \gamma\,c\,\mathcal{E}_{\mathsf{G}}^{X_2}(t) + f_j^*\left(\frac{t}{2}\right)$$

uniformly for all t, $0 < t < \delta$. Assuming that the simple functions are dense in X_2, we thus have the counterpart of (11.65), leading to the same kind of contradiction as above. □

Consequently the corresponding embedding

$$\mathrm{id} : X_1(U) \longrightarrow X_2(U)$$

can only be compact when

$$\liminf_{t\downarrow 0} \,\mathsf{q}_{\mathsf{G}}^{(X_1,X_2)}(t) = 0, \quad \text{or} \quad \liminf_{t\downarrow 0} \,\mathsf{q}_{\mathsf{C}}^{(X_1,X_2)}(t) = 0. \tag{11.66}$$

Another more direct approach between envelope results and compactness studies may be offered by estimates of entropy numbers and approximation numbers, respectively, in terms of moduli of continuity. We briefly recall these concepts.

Let A_1 and A_2 be two complex (quasi-) Banach spaces and let T be a linear and continuous operator from A_1 into A_2. If T is compact then for any given $\varepsilon > 0$ there are finitely many balls in A_2 of radius ε which cover the image TU_1 of the unit ball $U_1 = \{a \in A_1 : \|a|A_1\| \leq 1\}$.

Definition 11.14 *Let* A_1 *and* A_2 *be two complex (quasi-) Banach spaces,* $k \in \mathbb{N}$ *and let* $T : A_1 \to A_2$ *be a linear and continuous operator from* A_1 *into* A_2.

(i) *The kth entropy number* e_k *of* T *is the infimum of all numbers* $\varepsilon > 0$ *such that there exist* 2^{k-1} *balls in* A_2 *of radius* ε *which cover* TU_1.

(ii) *The kth approximation number* a_k *of* T *is the infimum of all numbers* $\|T - S\|$ *where* S *runs through the collection of all continuous linear maps from* A_1 *to* A_2 *with* rank $S < k$,

$$a_k(T) = \inf\{\|T - S\| : S \in \mathcal{L}(A_1, A_2),\ \text{rank } S < k\}.$$

For details and properties of approximation numbers we refer to [CS90], [EE87], [Kön86] and [Pie87] (restricted to the case of Banach spaces), and [ET96] for some extensions to quasi-Banach spaces. Among other features we only want to mention the multiplicativity of entropy numbers and approximation numbers: let A_1, A_2 and A_3 be complex (quasi-) Banach spaces and $T_1 : A_1 \longrightarrow A_2$, $T_2 : A_2 \longrightarrow A_3$ two operators in the sense of Definition 11.14. Then

$$e_{k_1+k_2-1}(T_2 \circ T_1) \le e_{k_1}(T_1)\, e_{k_2}(T_2), \quad k_1, k_2 \in \mathbb{N}, \tag{11.67}$$

and

$$a_{k_1+k_2-1}(T_2 \circ T_1) \le a_{k_1}(T_1)\, a_{k_2}(T_2), \quad k_1, k_2 \in \mathbb{N}. \tag{11.68}$$

Note that one has in general

$$\lim_{k \to \infty} e_k(T) = 0 \quad \text{if, and only if,} \quad T \quad \text{is compact .} \tag{11.69}$$

The last equivalence justifies the saying that entropy numbers measure "how compact" an operator acts. This is one reason to study the asymptotic behaviour of entropy numbers (that is, their decay) for compact operators in detail. Dealing with approximation numbers there is no complete counterpart of (11.69) in general; in particular, one only has

$$\lim_{n \to \infty} a_n(T) = 0 \quad \text{implies} \quad T \text{ is compact.} \tag{11.70}$$

It is known that it may happen that $\lim_{n \to \infty} a_n(T) = \alpha(T) > 0$ for some compact $T \in \mathcal{L}(A, B)$ when B fails to have the approximation property, see [EE87] and [Pie80, Prop. 10.1.3, 10.1.4]. It is known that many spaces, like Hilbert spaces, L_p, $1 \le p \le \infty$, $C(K)$, and ℓ_p, $1 \le p < \infty$, possess this property, but there exist spaces without it, cf. [Enf73], [Pie80, Thm. 10.4.7].

Moreover, approximation numbers – unlike entropy numbers – can be regarded as special *s-numbers*, a concept introduced by Pietsch [Pie80, Sect. 11].

Example 11.15 Let $m \in \mathbb{N}$, and X a real Banach space with $\dim X = m < \infty$, $\mathrm{id}_X \in \mathcal{L}(X, X)$ the natural embedding map. Then

$$2^{-\frac{k-1}{m}} \le e_k(\mathrm{id}_X) \le 4\, 2^{-\frac{k-1}{m}}, \quad k \in \mathbb{N},$$

and

$$a_k(\mathrm{id}_X) = \begin{cases} 1 , k = 1, \dots, m, \\ 0 , k > m. \end{cases}$$

In general, one can show for real Banach spaces X and Y, and $T \in \mathcal{L}(X, Y)$, that

$$\mathrm{rank}\, T = m \quad \text{if, and only if,}$$

$$\exists\, c > 0 \quad \forall\, n \in \mathbb{N} : c\, 2^{-\frac{n-1}{m}} \le e_n(T) \le 4\|T\| 2^{-\frac{n-1}{m}}, \tag{11.71}$$

see [EE87, Prop. II.1.3], [CS90, Prop. 1.3.1]. The counterpart for approximation numbers reads as

$$a_k(T) = 0 \quad \text{if, and only if,} \quad \text{rank } T < k. \tag{11.72}$$

Example 11.16 We briefly mention a second famous example, the *diagonal operator* $D : \ell_p \longrightarrow \ell_p$. Let $(\sigma_k)_k$ be a monotonically decreasing sequence, $\sigma_1 \geq \sigma_2 \geq \cdots \geq 0$, and consider

$$D_\sigma : \ell_p \longrightarrow \ell_p, \quad D_\sigma : x = (\xi_k)_k \longmapsto (\sigma_k \xi_k)_k, \quad 1 \leq p \leq \infty.$$

For convenience, let ℓ_p be real; then

$$a_k(D_\sigma) = \sigma_k, \quad k \in \mathbb{N},$$

and

$$\sup_{m \in \mathbb{N}} \; 2^{-\frac{k-1}{m}} \, (\sigma_1 \cdots \sigma_m)^{\frac{1}{m}} \; \leq \; e_k(D_\sigma) \; \leq \; 6 \sup_{m \in \mathbb{N}} \; 2^{-\frac{k-1}{m}} \, (\sigma_1 \cdots \sigma_m)^{\frac{1}{m}} \, ,$$

see [Pie80, Thms. 11.3.2, 11.11.4], [Kön86], [GKS87], [CS90, Prop. 1.3.2].

Remark 11.17 In view of the above simple examples one might be tempted to find a general relation between entropy numbers and approximation numbers. Obviously, Example 11.15 implies that an estimate of the type $e_k(T) \leq c \, a_k(T)$, $k \in \mathbb{N}$, cannot hold in general, whereas it is, for instance, always true that $\lim_{k \to \infty} e_k(T) \leq a_m(T)$, $m \in \mathbb{N}$; cf. [CS90, Lemma 2.5.2]. Clearly, the converse inequality, $a_m(T) \leq c \, e_m(T)$ – though being true in the context of (real) Hilbert spaces, [CS90, Thm. 3.4.2] – cannot hold in general either, since there are compact operators T with $\lim_{k \to \infty} a_k(T) > 0$ (if the target space fails to have the approximation property), but $\lim_{k \to \infty} e_k(T) = 0$ by (11.69).
However, replacing the term-wise estimates by particularly weighted averages, one obtains final answers of the following type: Let $0 < r < \infty$, X, Y Banach spaces, and $T \in \mathcal{L}(X, Y)$, then there exists a constant $c = c(r) > 0$, such that for $m \in \mathbb{N}$,

$$\sup_{k=1,\ldots,m} k^{\frac{1}{r}} \, e_k(T) \; \leq \; c \sup_{k=1,\ldots,m} k^{\frac{1}{r}} \, a_k(T),$$

see [Car81], [CS90, Thm. 3.1.1]. There are parallel results in [CS90, Thm. 3.1.2] refining the above ℓ_∞-setting. Another extension was found by Triebel in [Tri94]: There exists $c > 0$ such that for all $k \in \mathbb{N}$, $e_k(T) \leq c \, a_k(T)$, assuming that there is some $c' > 0$ with $a_{2j-1}(T) \leq c' \, a_{2j}(T)$ for all $j \in \mathbb{N}$, and $T \in \mathcal{L}(A, B)$ is compact. More generally, if there is a positive increasing function f on $\mathbb{N}$ with $f(2^j) \leq c f(2^{j-1})$ for some $c > 0$ and all $j \in \mathbb{N}$, this implies the existence of some $C > 0$ such that for all $k \in \mathbb{N}$,

$$\sup_{1 \leq j \leq k} f(j) \, e_j(T) \leq C \sup_{1 \leq j \leq k} f(j) \, a_j(T). \tag{11.73}$$

The interplay between continuity envelopes and approximation numbers relies on the following outcome.

Corollary 11.18 *Let $X(U)$ be a Banach space with $X(U) \hookrightarrow C(U)$. There exists $c > 0$ such that for all $k \in \mathbb{N}$,*

$$a_{k+1} \, (\mathrm{id} : X(U) \longrightarrow C(U)) \; \leq \; c \, k^{-\frac{1}{n}} \, \mathcal{E}_{\mathsf{C}}^X \left(k^{-\frac{1}{n}} \right). \tag{11.74}$$

Proof: We apply the following estimate obtained by Carl and Stephani in [CS90, Thm. 5.6.1]: Let (Y, d) be a compact metric space, X an arbitrary Banach space, and $T : X \to C(Y)$ compact, then

$$a_{k+1}(T) \leq \sup_{\|f|X\| \leq 1} \omega \left(Tf, \varepsilon_k(Y) \right), \quad k \in \mathbb{N}, \tag{11.75}$$

where $\varepsilon_k(Y)$ are the usual (non-dyadic) entropy numbers, i.e., the infimum of all numbers $\varepsilon > 0$ such that there exist $m \leq k$ balls of radius ε which cover Y. Adapted to our setting, $Y = \overline{U} \subset \mathbb{R}^n$, $T = \mathrm{id}$, the result follows immediately from Definition 5.1 of $\mathcal{E}_{\mathsf{C}}^X$, taking $\varepsilon_k(\overline{U}) \sim k^{-1/n}$ into account, $k \in \mathbb{N}$. $\qquad \square$

We combine Corollaries 11.18 and 11.11(i) and adapt the notation (11.48) to our setting: Let $X(U)$ be some function space of regular distributions with $X(U) \not\hookrightarrow L_\infty(U)$. Let $X^\nabla(U) = \{g \in D'(U) : D^\alpha g \in X(U), \, |\alpha| \leq 1\}$ be the subspace of $X(U)$ normed by

$$\big\| g | X^\nabla(U) \big\| \; = \; \sum_{|\alpha| \leq 1} \| D^\alpha g | X(U) \|.$$

Let $C(U)$ stand for the space of all complex-valued bounded uniformly continuous functions on $\overline{U}$, equipped with the sup-norm as usual.

Corollary 11.19 *Let $X(U), X^\nabla(U)$ be given with $X(U) \hookrightarrow B_{\infty,\infty}^{-1}(U)$ and $X^\nabla(U) \hookrightarrow C(U)$. Let $\mathcal{E}_{\mathsf{G}}^X$ satisfy (11.58), and assume that there is a bounded (linear) lift operator L mapping $X(U)$ into $X^\nabla(U)$ such that its inverse L^{-1} exists and maps $C(U)$ into $B_{\infty,\infty}^{-1}(U)$. Then there is some $c > 0$ such that*

$$a_k \left(\mathrm{id}_X : X(U) \longrightarrow B_{\infty,\infty}^{-1}(U) \right) \; \leq \; c \, k^{-\frac{1}{n}} \, \mathcal{E}_{\mathsf{G}}^X \left(k^{-1} \right), \quad k \in \mathbb{N}. \tag{11.76}$$

Proof: We combine Corollary 11.18 for compact embeddings in $C(U)$ (as target space) with the properties of the operator L and its inverse L^{-1} and Corollary 11.11(i) (which requires (11.58)); in particular, using the bounded lift $L : X(U) \longrightarrow X^\nabla(U)$ with $L^{-1} : B^0_{\infty,\infty}(U) \longrightarrow B^{-1}_{\infty,\infty}(U)$, the decomposition

$$\mathrm{id}_X = L^{-1} \circ \left(C(U) \hookrightarrow B^0_{\infty,\infty}(U) \right) \circ \left(\mathrm{id}_{X^\nabla} : X^\nabla(U) \longrightarrow C(U) \right) \circ L,$$

an application of Corollary 11.11(i) for id_{X^∇}, together with the multiplicativity of approximation numbers conclude the argument. $\square$

Dealing with entropy numbers, we conclude from (11.73) and Corollaries 11.18, 11.19 the following result.

Corollary 11.20 *Let* $f : \mathbb{N} \to \mathbb{R}$ *be a positive and increasing function satisfying*

$$f\left(2^k\right) \leq c\, f\left(2^{k-1}\right) \tag{11.77}$$

for some $c > 0$ *and all* $k \in \mathbb{N}$.

(i) *Let* $X(U)$ *be a Banach space with* $X(U) \hookrightarrow C(U)$. *There exists* $C > 0$ *such that for all* $m \in \mathbb{N}$,

$$\sup_{1 \leq k \leq m} f(k)\, e_k\, (\mathrm{id} : X(U) \longrightarrow C(U))$$

$$\leq C \sup_{1 \leq k \leq m} f(k)\, k^{-\frac{1}{n}}\, \mathcal{E}^X_C\left(k^{-\frac{1}{n}} \right). \tag{11.78}$$

(ii) *Let* $X(U), X^\nabla(U)$ *be given with* $X(U) \hookrightarrow B^{-1}_{\infty,\infty}(U)$ *and* $X^\nabla(U) \hookrightarrow C(U)$. *Let* $\mathcal{E}^X_G$ *satisfy (11.58), and assume that there is a bounded (linear) lift operator* L *mapping* $X(U)$ *into* $X^\nabla(U)$ *such that its inverse* L^{-1} *exists and maps* $C(U)$ *into* $B^{-1}_{\infty,\infty}(U)$. *Then there is some* $C > 0$ *such that for all* $m \in \mathbb{N}$,

$$\sup_{1 \leq k \leq m} f(k)\, e_k \left(\mathrm{id} : X(U) \longrightarrow B^{-1}_{\infty,\infty}(U) \right)$$

$$\leq C \sup_{1 \leq k \leq m} f(k)\, k^{-\frac{1}{n}}\, \mathcal{E}^X_G\left(k^{-1} \right). \tag{11.79}$$

Obviously, (11.74) and (11.76) only provide upper estimates for the corresponding approximation numbers; we shall discuss the sharpness of these bounds in different settings, and start with a short account on what is known for spaces of type $A^s_{p,q}$. Let $-\infty < s_2 \leq s_1 < \infty$, $0 < p_1, p_2 \leq \infty$ ($p_1, p_2 < \infty$ in the F-case), $0 < q_1, q_2 \leq \infty$, and

$$\mathrm{id}_A = \mathrm{id} : A^{s_1}_{p_1,q_1}(U) \longrightarrow A^{s_2}_{p_2,q_2}(U),$$

where the spaces $A^s_{p,q}(U)$ are given by (7.24). Then id_A is continuous when

$$\delta_+ := s_1 - s_2 - n \left(\frac{1}{p_1} - \frac{1}{p_2} \right)_+ \geq 0 \qquad (11.80)$$

and $q_1 \leq q_2$ if $\delta_+ = 0$ in the B-case. Furthermore, id_A becomes compact when $\delta_+ > 0$; cf. [ET96, (2.5.1/10)]. The extension to values $p_2 < p_1$ – compared with the $\mathbb{R}^n$- setting – is due to Hölder's inequality.

In this situation Edmunds and Triebel proved in [ET89], [ET92] (see also [ET96, Thm. 3.3.3/2]) that

$$e_k(\mathrm{id}_A) \sim k^{-\frac{s_1 - s_2}{n}}, \qquad k \in \mathbb{N}, \qquad (11.81)$$

where $s_1 \geq s_2$, $0 < p_1, p_2 \leq \infty$ ($p_1, p_2 < \infty$ in the F-case), $0 < q_1, q_2 \leq \infty$, and $\delta_+ > 0$. In the case of approximation numbers the situation is more complicated; the result of Edmunds and Triebel in [ET96, Thm. 3.3.4], partly improved by Caetano [Cae98] and Skrzypczak [Skr05] reads as

$$a_k(\mathrm{id}_A) \sim k^{-\frac{\delta_+}{n} - \varkappa}, \qquad k \in \mathbb{N}, \qquad (11.82)$$

with

$$\varkappa = \left(\frac{\min(p'_1, p_2)}{2} - 1 \right)_+ \cdot \min \left(\frac{\delta}{n}, \frac{1}{\min(p'_1, p_2)} \right), \qquad (11.83)$$

and δ is given by (7.34). The additional exponent $\varkappa$ only appears when $p_1 < 2 < p_2$. The above asymptotic result is almost complete now, apart from the restriction that $(p_1, p_2) \neq (1, \infty)$ when $0 < p_1 < 2 < p_2 \leq \infty$.

In particular, when $p_2 = \infty$, $s_1 \geq s_2$, $0 < p_1 \leq \infty$, $0 < q_1, q_2 \leq \infty$, and

$$\delta_+ = \delta = s_1 - s_2 - \frac{n}{p_1} > 0,$$

then

$$e_k \left(\mathrm{id} : B^{s_1}_{p_1, q_1}(U) \longrightarrow B^{s_2}_{\infty, q_2}(U) \right) \sim k^{-\frac{s_1 - s_2}{n}}, \qquad k \in \mathbb{N}, \qquad (11.84)$$

and

$$a_k \left(\mathrm{id} : B^{s_1}_{p_1, q_1}(U) \longrightarrow B^{s_2}_{\infty, q_2}(U) \right)$$

$$\sim \begin{cases} k^{-\frac{s_1 - s_2}{n} + \frac{1}{p_1}} & , 2 \leq p_1 \leq \infty \\ k^{-\frac{s_1 - s_2}{n} + \frac{1}{2}} & , 1 < p_1 < 2, \ s_1 - s_2 > n \\ k^{-\left[\frac{s_1 - s_2}{n} - \frac{1}{p_1} \right] \frac{p'_1}{2}} & , 1 < p_1 < 2, \ s_1 - s_2 \leq n \end{cases}, \qquad (11.85)$$

(and two-sided estimates for $0 < p_1 \leq 1$); see [ET96, Thm. 3.3.4], [Cae98], [Skr05].

Remark 11.21 In [HM04] and [CH05] we studied entropy numbers and approximation numbers, respectively, for compact embeddings of spaces of generalised smoothness. Restricted to our Example 7.5, $s \in \mathbb{R}$, $b \in \mathbb{R}$, $0 < p < \infty$, $0 < q \leq \infty$, with $\frac{n}{p} < s \leq \frac{n}{p} + 1$, the compactness is an immediate consequence of the above argument and (7.62). In [HM06, Prop. 4.4, Example 4.8] we proved that for $k \in \mathbb{N}$,

$$a_k \left(\mathrm{id} : B_{p,q}^{s_1,b}(U) \longrightarrow B_{\infty,\infty}^{s_2}(U) \right) \sim k^{-\frac{s_1-s_2}{n}+\frac{1}{p}} (1+\log k)^{-b}, \qquad (11.86)$$

assuming $2 \leq p \leq \infty$ for convenience, and $0 < q \leq \infty$, $\frac{n}{p} < s_1 - s_2 < \frac{n}{p} + 1$, $b \in \mathbb{R}$. For $s_2 = 0$ the target space can be replaced by $C(U)$. In the case of $s = \frac{n}{p} + 1$, $2 \leq p \leq \infty$, we obtained in [CH05, Prop. 4.10],

$$c_1 \, k^{-\frac{1}{n}} (1+\log k)^{-b} \leq a_k \left(\mathrm{id} : B_{p,q}^{1+n/p,b}(U) \longrightarrow B_{\infty,\infty}^{s_2}(U) \right)$$

$$\leq c_2 \, k^{-\frac{1}{n}} \left\{ \begin{array}{ll} (1+\log k)^{\frac{1}{q'}-b} & , b < \frac{1}{q'} \\ (\log (1+\log k))^{\frac{1}{q'}} & , b = \frac{1}{q'} \end{array} \right\},$$

for $k \in \mathbb{N}$, see also Example 9.8.

The counterpart for entropy numbers, can be found in [Leo00, Thm. 3],

$$e_k \left(\mathrm{id} : B_{p,q}^{s_1,b}(U) \longrightarrow B_{\infty,\infty}^{s_2}(U) \right) \sim k^{-\frac{s_1-s_2}{n}} (1+\log k)^{-b}, \qquad (11.87)$$

where $0 < p, q \leq \infty$, $s_1 - s_2 > \frac{n}{p}$, $b \in \mathbb{R}$, and $k \in \mathbb{N}$.

We consider the natural embedding operators

$$\mathrm{id}_X^1 : X(U) \longrightarrow C(U), \qquad (11.88)$$

and

$$\mathrm{id}_X^2 : X(U) \longrightarrow B_{\infty,\infty}^{-1}(U), \qquad (11.89)$$

where the spaces $X(U)$ are defined by restriction.

Remark 11.22 One can also introduce spaces of type $\mathrm{Lip}_{p,q}^{(1,-\alpha)}(U)$, or $C(U)$, by the usual adaption of the corresponding definitions, e.g., $\mathrm{Lip}^{(1,-\alpha)}(U)$ as the set of those $f \in C(\overline{U})$ such that

$$\left\| f | \mathrm{Lip}^{(1,-\alpha)}(U) \right\| = \| f | L_\infty(U) \| + \sup_{\substack{x \in U, h \in \mathbb{R}^n \\ 0 < |h| < 1/2}} \frac{|(\Delta_h f)(x)|}{|h| \, |\log |h||^\alpha} \qquad (11.90)$$

is finite. Standard procedures show that there is a bounded extension map from $X(U)$ to $X(\mathbb{R}^n)$ in these cases; see, for example, [EE87, pp. 250-251].

We discuss a few cases for X and begin with special settings for (11.88).

Example 11.23 Let $X = A^s_{p,q}$ with

$$\frac{n}{p} < s < \frac{n}{p} + 1, \quad 0 < p \leq \infty, \quad 0 < q \leq \infty,$$

or

$$s = \frac{n}{p} + 1, \quad 0 < p \leq \infty, \quad 1 < q \leq \infty.$$

For convenience, we may restrict ourselves to B-spaces only, though the F-case can be handled completely parallel. Note that – in view of (7.31) – the above assumption $\delta_+ > 0$ for compactness implies $s > \frac{n}{p}$, $0 < q \leq \infty$; see also our remarks in Section 7.2 or Proposition 11.13(ii) (together with the corresponding results in the previous sections). Obviously,

$$\mathrm{id}^1_{B^s_{p,q}} : B^s_{p,q}(U) \longrightarrow C(U)$$

remains compact for $s > \frac{n}{p} + 1$, but our envelope concept is not adapted appropriately for this higher smoothness; so this loss of information causes very weak estimates only (and will not be discussed further).

We conclude from our results in Section 9.1 and (11.74), (11.78),

$$e_k\left(\mathrm{id}^1_{B^s_{p,q}}\right) \leq c\, a_k\left(\mathrm{id}^1_{B^s_{p,q}}\right)$$
$$\leq c' \begin{cases} k^{-\frac{s}{n}+\frac{1}{p}} & , \frac{n}{p} < s < \frac{n}{p}+1\,, 0 < q \leq \infty \\ k^{-\frac{1}{n}}(\log\langle k\rangle)^{\frac{1}{q'}} & , \quad s = \frac{n}{p}+1\,, 1 < q \leq \infty \end{cases}.$$

We compare this result with (11.84), (11.85) (with $s_2 = 0$) and realise that for $0 < s - \frac{n}{p} < 1$ (i.e., in the *"super-critical strip"*) we are led to the correct upper estimates for $a_k(\mathrm{id}^1_{B^s_{p,q}})$ apart from the case $1 < p < 2$, whereas otherwise – as well as for entropy numbers – our method provides a less sharp upper bound only. This, however, is not very surprising, as, firstly, the direct link is given between approximation numbers and envelopes (hence the entropy numbers being only some by-product in that sense), and, secondly, our continuity envelope functions are "made" for $0 \leq s - \frac{n}{p} \leq 1$ only; otherwise they lack some interesting information. The more astonishing observation in our opinion is rather the sharpness of the results otherwise.

Example 11.24 As a second case for X from (11.88) we regard Lipschitz spaces, $X = \mathrm{Lip}^{(1,-\alpha)}_{p,q}$. We begin with $X(U) = \mathrm{Lip}^{(1,-\alpha)}(U)$, $\alpha > 0$. The compactness of

$$\mathrm{id}^1_{\mathrm{Lip}} : \mathrm{Lip}^{(1,-\alpha)}(U) \longrightarrow C(U)$$

is a consequence of [EH00, Cor. 3.19]. Now Theorem 6.6(ii) and (11.74) yield

$$e_k\left(\mathrm{id}_{\mathrm{Lip}}^1\right) \;\leq\; c\,a_k\left(\mathrm{id}_{\mathrm{Lip}}^1\right) \;\leq\; c'\,k^{-\frac{1}{n}}\,(\log\langle k\rangle)^{\alpha},$$

which by [EH00, Cor. 3.19(i)] gives the exact asymptotic behaviour both for entropy numbers and approximation numbers.

Assume $0 < q \leq \infty$, $n < p \leq \infty$, $\alpha > \frac{1}{q}$, then Corollary 9.9 and (11.74) imply for

$$\mathrm{id}^1_{\mathrm{Lip}_{p,q}^{(1,-\alpha)}} \;:\; \mathrm{Lip}_{p,q}^{(1,-\alpha)}(U) \longrightarrow C(U)$$

and $k \in \mathbb{N}$,

$$e_k\left(\mathrm{id}^1_{\mathrm{Lip}_{p,q}^{(1,-\alpha)}}\right) \;\leq\; c\,a_k\left(\mathrm{id}^1_{\mathrm{Lip}_{p,q}^{(1,-\alpha)}}\right) \;\leq\; c'\,k^{-\frac{1}{n}+\frac{1}{p}}\,(\log\langle k\rangle)^{\alpha-\frac{1}{q}}.$$

In view of (9.34) and (11.86) this is the precise asymptotic description,

$$a_k\left(\mathrm{id}^1_{\mathrm{Lip}_{p,q}^{(1,-\alpha)}}\right) \;\sim\; k^{-\frac{1}{n}+\frac{1}{p}}\,(\log\langle k\rangle)^{\alpha-\frac{1}{q}}, \quad k \in \mathbb{N}, \tag{11.91}$$

with $0 < q \leq \infty$, $n < p \leq \infty$, $\alpha > \frac{1}{q}$, whereas (11.87) and (9.34) lead to better estimates for entropy numbers in that case,

$$e_k\left(\mathrm{id}^1_{\mathrm{Lip}_{p,q}^{(1,-\alpha)}}\right) \;\sim\; k^{-\frac{1}{n}}\,(\log\langle k\rangle)^{\alpha-\frac{1}{q}}, \quad k \in \mathbb{N}. \tag{11.92}$$

We consider some cases for X in (11.89).

Example 11.25 Let again $X = B_{p,q}^s$, now with

$$\left\{ \begin{array}{ll} \frac{n}{p} - 1 < s < \frac{n}{p}, & 0 < p \leq n,\ 0 < q \leq \infty \\[4pt] \sigma_p < s < \frac{n}{p}, & n < p \leq \infty,\ 0 < q \leq \infty \\[4pt] s = \frac{n}{p}, & 0 < p < \infty,\ 1 < q \leq \infty \\[4pt] s = 0, & n < p < \infty,\ 0 < q \leq \min(p,2) \end{array} \right\}.$$

Similar to our above remarks, the compactness assumption $\delta_+ > 0$ excludes $s \leq \frac{n}{p} - 1$, whereas $s > \frac{n}{p}$ is omitted because of (necessarily) weaker estimates using (inappropriately adapted) growth envelope techniques.

Dealing with the (sub-) critical case we are led to

$$a_k\left(\mathrm{id}^2_{B_{p,q}^s}\right) \;\leq\; c\,k^{-\frac{s+1}{n}+\frac{1}{p}}, \quad k \in \mathbb{N},$$

if $\sigma_p < s < \frac{n}{p}$, $0 < q \le \infty$, or $s = 0$, $1 < p < \infty$, $0 < q \le \min(p,2)$ by our results in Section 8.1, 8.3 and (11.76); recall notation (7.2). Comparison with (11.85) confirms the sharpness in case of $\sigma_p < s < \frac{n}{p}$, $n \ge 2$; otherwise we can repeat our above discussion. This argument applies to entropy numbers, too. Note that the existence of the lift operator L can be verified applying usual restriction-extension procedures and the lift operator I_σ in $\mathbb{R}^n$ given by (7.6), which maps $B_{p,q}^s$ isomorphically onto $B_{p,q}^{s-\sigma}$ for all admitted parameters. Alternatively one can also use regular elliptic differential operators adapted to U; see [Tri78a, Thm. 4.9.2] for the case $1 < p < \infty$, $1 \le q \le \infty$, and [Tri83, Thm. 4.3.4] for the extensions to $0 < p,q \le \infty$, which are based on more recent techniques of Fourier multipliers.

Example 11.26 Finally, we consider $X = L_p(\log L)_a$, with

$$\begin{cases} n < p < \infty, a \in \mathbb{R} \\ p = n, \quad a > 0 \\ p = \infty, \quad a \le 0 \end{cases}$$

The compactness of

$$\mathrm{id}_{L_p(\log L)_a}^2 \; : \; L_p(\log L)_a(U) \longrightarrow B_{\infty,\infty}^{-1}(U)$$

is confirmed by [ET96, (2.5.1/10), Props. 2.6.1/1,2, Thm. 3.4.3/1] together with (11.80) and a duality argument. The existence of a bounded linear lift is covered by [ET96, Thm. 2.6.3], at least for $n \le p < \infty$. Theorem 4.7(ii) combined with (11.53) for $\mu = \frac{1}{p}$, $\varkappa = -a$, and (11.76) provides

$$a_k(\mathrm{id}_{L_p(\log L)_a}^2) \le c \begin{cases} k^{-\frac{1}{n}+\frac{1}{p}} (\log\langle k \rangle)^{-a} , \; n < p < \infty , \; a \in \mathbb{R} \\ (\log\langle k \rangle)^{-a} \quad\quad , \quad\quad p = n \quad , \; a > 1 \end{cases}. \quad (11.93)$$

This observation (11.93) led us in [CH03] to the sharp asymptotic estimate:

$$a_k \left(\mathrm{id} : L_p(\log L)_a(U) \longrightarrow B_{\infty,\infty}^{-1}(U) \right) \sim k^{-\frac{1}{n}+\frac{1}{p}} (\log\langle k \rangle)^{-a} ,$$

for $k \in \mathbb{N}$, where $a > 0$, $n < p < \infty$ for $n \ge 2$, and $2 \le p < \infty$ when $n = 1$. The additional restriction $p \ge 2$ is caused by the lower estimate, but was somehow to be expected in the case of approximation numbers, see (11.85). Instead of Corollary 11.20(ii) and Theorem 4.7(ii) we apply (11.84) and $L_{p+\varepsilon}(U) \hookrightarrow L_p(\log L)_a(U) \hookrightarrow L_p(U)$, $\varepsilon > 0$, $a > 0$, and conclude

$$e_k \left(\mathrm{id} : L_p(\log L)_a(U) \longrightarrow B_{\infty,\infty}^{-1}(U) \right) \sim k^{-\frac{1}{n}}, \quad k \in \mathbb{N},$$

for $a > 0$, $n < p < \infty$. As for the limiting case $p = n$, there are entropy number results in [ET96, Sect. 3.4], [EN98], [Cae00], but – as far as we know – no complete results for approximation numbers yet.

Remark 11.27 It is clear that the above method can be applied to a lot of situations whenever envelope results are at hand already, e.g., certain weighted spaces, spaces of generalised smoothness, etc. Usually, by the arguments explicated above, one would expect sharp asymptotic results for approximation numbers (at least in suitably adapted situations), that can be transferred to "dual" embedding operators without difficulties. Concerning entropy numbers the present approach seems too weak in general.

Remark 11.28 The study of entropy numbers and approximation numbers of embeddings between function spaces is closely related to the distribution of eigenvalues of (degenerate) elliptic operators, as the books [ET96] and [Tri97] show. We conclude this section with a brief description of the background and the context for some further possible applications of our results.

The motivation comes from Carl's inequality giving an excellent link to possible applications, in particular, between entropy numbers and eigenvalues of some compact operators. The setting is the following. Let A be a complex (quasi-) Banach space and $T \in \mathcal{L}(A)$ compact. Then the spectrum of T (apart from the point 0) consists only of eigenvalues of finite algebraic multiplicity. Let $\{\mu_k(T)\}_{k \in \mathbb{N}}$ be the sequence of all non-zero eigenvalues of T, repeated according to algebraic multiplicity and ordered such that

$$|\mu_1(T)| \geq |\mu_2(T)| \geq \cdots \geq 0.$$

Then Carl's inequality states that

$$\left(\prod_{m=1}^{k} |\mu_m(T)| \right)^{1/k} \leq \inf_{n \in \mathbb{N}} 2^{\frac{n}{2k}} e_n(T), \qquad k \in \mathbb{N}.$$

In particular, we have

$$|\mu_k(T)| \leq \sqrt{2}\, e_k(T). \tag{11.94}$$

This result was originally proved by Carl in [Car81] and Carl and Triebel in [CT80] when A is a Banach space. An extension to quasi-Banach spaces is given in [ET96, Thm. 1.3.4]. When A is a Banach space, Zemánek [Zem80] could prove

$$\lim_{m \to \infty} (e_k(T^m))^{\frac{1}{m}} = r(T), \quad k \in \mathbb{N},$$

where $r(T)$ is the spectral radius of T, see also [EE87, Cor. II.1.7].

Concerning estimates from below, i.e., converse to (11.94), and the connection between approximation numbers and eigenvalues, it is reasonable to concentrate on the Hilbert space setting first. Let H be a complex Hilbert space and $T \in \mathcal{L}(H)$ compact, the non-zero eigenvalues of which are denoted by $\{\mu_k(T)\}_{k \in \mathbb{N}}$ again; then T^*T has a non-negative, self-adjoint, compact square root $|T|$, and for all $k \in \mathbb{N}$,

$$a_k(T) = \mu_k(|T|), \tag{11.95}$$

see [EE87, Thm. II.5.10]. Hence, if in addition T is non-negative and self-adjoint, then the approximation numbers of T coincide with its eigenvalues. Moreover, a famous inequality of Weyl, see [Kön86, Thm. 1.b.5], states that for all $n \in \mathbb{N}$,

$$\prod_{j=1}^{n} |\mu_j(T)| \leq \prod_{j=1}^{n} a_j(T),$$

from which it follows that for all $n \in \mathbb{N}$ and all $p \in (0, \infty)$,

$$\sum_{j=1}^{n} |\mu_j(T)|^p \leq \sum_{j=1}^{n} a_j^p(T).$$

Outside Hilbert spaces the results are less good but still very interesting. Let A be a complex Banach space, $T \in \mathcal{L}(A)$ compact, and $\{\mu_k(T)\}_{k \in \mathbb{N}}$ its eigenvalue sequence. König proved that for all $m \in \mathbb{N}$ and all $p \in (0, \infty)$,

$$\left(\sum_{j=1}^{m} |\mu_j(T)|^p \right)^{\frac{1}{p}} \leq K_p \left(\sum_{j=1}^{m} a_j^p(T) \right)^{\frac{1}{p}},$$

where $K_p = 2e/\sqrt{p}$ if $0 < p < 1$, and $K_p = 2^{1/p}\sqrt{2e}$, if $1 \leq p < \infty$. For details and further remarks we refer to [CS90], [EE87], [Kön86] and [Pie87].

In view of Carl's inequality (11.94) and (11.95) one may thus obtain upper and lower estimates for eigenvalues from the study of entropy and approximation numbers, respectively, at least in Hilbert space settings when $T \in \mathcal{L}(H)$ is compact, non-negative and self-adjoint, see (11.95). Sometimes one can furthermore prove that the root spaces coincide and then – by some tricky bootstrapping techniques – "shift" estimates (originally proved in Hilbert spaces) to (quasi-) Banach spaces. The problem to determine $e_k(T)$ or $a_k(T)$, respectively, can often be reduced further to the study of entropy numbers or approximation numbers of suitable embeddings assuming that one has corresponding Hölder inequalities available.

Another possible application is connected with the so-called "negative spectrum" and the Birman-Schwinger principle as described in [Sch86, Ch. 8, Sect. 5]. Let A be a self-adjoint operator acting in a Hilbert space H and let A be positive. Let V be a closable operator acting in H and suppose that $K : H \to H$ is a compact linear operator such that

$$Ku = VA^{-1}V^*u \qquad \text{for all} \quad u \in \text{dom}(VA^{-1}V^*)$$

where V^* is the adjoint of V. Assume that $\text{dom}(A) \cap \text{dom}(V^*V)$ is dense in H. Then the above-mentioned result provides: $A - V^*V$ has a self-adjoint extension G with pure point spectrum in $(-\infty, 0]$ such that

$$\# \{\sigma(G) \cap (-\infty, 0] \} \leq \# \{k \in \mathbb{N} : |\lambda_k| \geq 1\}$$

where $\{\lambda_k\}$ is the sequence of eigenvalues of K, counted according to their multiplicity and ordered by decreasing modulus. In particular, we consider the behaviour of the "negative spectrum" $\sigma(G_\nu) \cap (-\infty, 0]$ of the self-adjoint unbounded operator

$$G_\nu = a(x, D) - \nu b^2(x) \qquad \text{as} \quad \nu \to \infty \tag{11.96}$$

where

$$a(x, D) \in \Psi^{\varkappa}_{1,\gamma}, \quad \varkappa > 0, \quad 0 \le \gamma < 1, \tag{11.97}$$

is assumed to be a positive-definite and self-adjoint pseudodifferential operator in L_2 and $b(x)$ is a real-valued function. We know from former considerations, cf. [HT94b, 2.4, 5.2], that

$$\#\{\sigma(G_\nu) \cap (-\infty, 0]\} \le \#\left\{k \in \mathbb{N} : \sqrt{2}\, e_k \ge \nu^{-1}\right\} \tag{11.98}$$

with $e_k = e_k\big(b(x)\, b(x, D)\, b(x)\big)$ and $b(x, D) = a^{-1}(x, D) \in \Psi^{-\varkappa}_{1,\gamma}$.

These are essentially the applications we have in mind for using our results on entropy numbers and approximation numbers of compact embeddings. This programme was carried out in [HT94b], [ET96], first, and [Tri97], [Har98], [Har00a], [EH00] in different settings afterwards; we refer to these papers and books for details.

References

[AF03]   R.A. Adams and J.J.F. Fournier. *Sobolev Spaces*, volume 140 of *Pure and Applied Mathematics*. Elsevier, Academic Press, 2nd edition, 2003. ▶ 4, 26, 27, 28

[Alv77a] A. Alvino. A limit case of the Sobolev inequality in Lorentz spaces. *Rend. Accad. Sci. Fis. Mat. Napoli (4)*, 44:105–112, 1977. Italian. ▶ 70

[Alv77b] A. Alvino. Sulla diseguaglianza di Sobolev in spazi di Lorentz. *Boll. Un. Mat. Ital. A (5)*, 14(1):148–156, 1977. Italian. ▶ 70

[BH06]   M. Bownik and K.P. Ho. Atomic and molecular decompositions of anisotropic Triebel-Lizorkin spaces. *Trans. Amer. Math. Soc.*, 358(4):1469–1510, 2006. ▶ 127

[BL76]   J. Bergh and J. Löfström. *Interpolation Spaces*. Springer, Berlin, 1976. ▶ 61, 70, 120, 137

[BL02]   G. Bourdaud and M. Lanza de Cristoforis. Functional calculus on Hölder-Zygmund spaces. *Trans. Amer. Math. Soc.*, 354(10):4109–4129, 2002. ▶ 155, 156

[BM03]   M. Bricchi and S.D. Moura. Complements on growth envelopes of spaces with generalised smoothness in the sub-critical case. *Z. Anal. Anwendungen*, 22(2):383–398, 2003. ▶ 116, 123

[Bow05] M. Bownik. Atomic and molecular decompositions of anisotropic Besov spaces. *Math. Z.*, 250(3):539–571, 2005. ▶ 127, 128, 176

[BPT96] H.-Q. Bui, M. Paluszyński, and M.H. Taibleson. A maximal function characterization of weighted Besov-Lipschitz and Triebel-Lizorkin spaces. *Studia Math.*, 119(3):219–246, 1996. ▶ 127

[BPT97] H.-Q. Bui, M. Paluszyński, and M.H. Taibleson. Characterization of the Besov-Lipschitz and Triebel-Lizorkin spaces. The case $q < 1$. *J. Fourier Anal. Appl.*, 3 (Spec. Iss.):837–846, 1997. ▶ 127

[BR80]   C. Bennett and K. Rudnick. On Lorentz-Zygmund spaces. *Dissertationes Math.*, 175:72 pp., 1980. ▶ 17, 18, 83, 137

[Bru72] Y.A. Brudnyi. On the rearrangement of a smooth function. *Uspechi Mat. Nauk*, 27:165–166, 1972. ▶ 123

[BS88] C. Bennett and R. Sharpley. *Interpolation of operators.* Academic Press, Boston, 1988. ▶ 11, 13, 15, 17, 19, 20, 21, 42, 52, 53, 54, 55, 78, 81, 102, 103, 121, 136, 143, 156, 162, 163

[Bui82] H.-Q. Bui. Weighted Besov and Triebel spaces: Interpolation by the real method. *Hiroshima Math. J.*, 12(3):581–605, 1982. ▶ 127, 175, 186

[Bui84] H.-Q. Bui. Characterizations of weighted Besov and Triebel-Lizorkin spaces via temperatures. *J. Funct. Anal.*, 55(1):39–62, 1984. ▶ 127

[BW80] H. Brézis and S. Wainger. A note on limiting cases of Sobolev embeddings and convolution inequalities. *Comm. Partial Differential Equations*, 5:773–789, 1980. ▶ ix, 4, 7, 35, 58, 113, 155

[Cae98] A. Caetano. About approximation numbers in function spaces. *J. Approx. Theory*, 94:383–395, 1998. ▶ 202

[Cae00] A. Caetano. Entropy numbers of embeddings between logarithmic Sobolev spaces. *Port. Math.*, 57(3):355–379, 2000. ▶ 206

[Car81] B. Carl. Entropy numbers, *s*-numbers and eigenvalue problems. *J. Funct. Anal.*, 41:290–306, 1981. ▶ 199, 207

[CF06] A.M. Caetano and W. Farkas. Local growth envelopes of Besov spaces of generalized smoothness. *Z. Anal. Anwendungen*, 25:265–298, 2006. ▶ 123, 142

[CH03] A.M. Caetano and D.D. Haroske. Sharp estimates of approximation numbers via growth envelopes. In: D.D. Haroske, Th. Runst, and H.J. Schmeißer, editors, *Function Spaces, Differential Operators and Nonlinear Analysis - The Hans Triebel Anniversary Volume*, pages 237–244. Birkhäuser, Basel, 2003. ▶ 206

[CH05] A.M. Caetano and D.D. Haroske. Continuity envelopes of spaces of generalised smoothness: A limiting case; embeddings and approximation numbers. *J. Funct. Spaces Appl.*, 3(1):33–71, 2005. ▶ 117, 156, 181, 183, 184, 203

[CH06] A.M. Caetano and D.D. Haroske. Growth envelopes for function spaces on fractal *h*-sets. Preprint, 2006. ▶ 8

[CL06] A.M. Caetano and H.-G. Leopold. Local growth envelopes of Triebel-Lizorkin spaces of generalized smoothness. Technical Report CM 06/I-3, University of Aveiro, Portugal, 2006. ▶ 123, 142

[CM04a] A.M. Caetano and S.D. Moura. Local growth envelopes of spaces of generalized smoothness: The critical case. *Math. Inequal. Appl.*, 7(4):573–606, 2004. ▶ 117, 142

[CM04b] A.M. Caetano and S.D. Moura. Local growth envelopes of spaces of generalized smoothness: The subcritical case. *Math. Nachr.*, 273:43–57, 2004. ▶ 123

[CP98] M. Cwikel and E. Pustylnik. Sobolev type embeddings in the limiting case. *J. Fourier Anal. Appl.*, 4:433–446, 1998. ▶ 4, 123

[CPSS01] M. Carro, L. Pick, J. Soria, and V.D. Stepanov. On embeddings between classical Lorentz spaces. *Math. Inequal. Appl.*, 4(3):397–428, 2001. ▶ 54

[CS90] B. Carl and I. Stephani. *Entropy, compactness and the approximation of operators.* Cambridge Univ. Press, Cambridge, 1990. ▶ 139, 198, 199, 200, 208

[CT80] B. Carl and H. Triebel. Inequalities between eigenvalues, entropy numbers and related quantities in Banach spaces. *Math. Ann.*, 251:129–133, 1980. ▶ 207

[DL93] R. DeVore and G.G. Lorentz. *Constructive Approximation*, volume 303 of *GMW*. Springer, Berlin, 1993. ▶ 13, 20, 21, 22, 76, 88, 102, 115

[DS84] R. DeVore and R. Sharpley. On the differentiability of functions in $\mathbb{R}^n$. *Proc. Amer. Math. Soc.*, 91(2):326–328, 1984. ▶ 80, 190

[EE87] D.E. Edmunds and W.D. Evans. *Spectral Theory and Differential Operators.* Clarendon Press, Oxford, 1987. ▶ 4, 26, 27, 28, 198, 199, 203, 207, 208

[EE04] D.E. Edmunds and W.D. Evans. *Hardy Operators, Function Spaces and Embeddings.* Springer Monographs in Mathematics. Springer, 2004. ▶ 4, 13

[EGO96] D.E. Edmunds, P. Gurka, and B. Opic. Sharpness of embeddings in logarithmic Bessel potential spaces. *Proc. R. Soc. Edinb., Sect. A*, 126(5):995–1009, 1996. ▶ 4

[EGO97] D.E. Edmunds, P. Gurka, and B. Opic. On embeddings of logarithmic Bessel potential spaces. *J. Funct. Anal.*, 146(1):116–150, 1997. ▶ 4, 114

[EGO00] D.E. Edmunds, P. Gurka, and B. Opic. Optimality of embeddings of logarithmic Bessel potential spaces. *Q. J. Math.*, 51(2):185–209, 2000. ▶ 4, 114

[EH99] D.E. Edmunds and D.D. Haroske. Spaces of Lipschitz type, embeddings and entropy numbers. *Dissertationes Math.*, 380:43 pp., 1999. ▶ 7, 22, 25, 112, 113, 114, 115, 156

[EH00] D.E. Edmunds and D.D. Haroske. Embeddings in spaces of Lipschitz type, entropy and approximation numbers, and applications. *J. Approx. Theory*, 104(2):226–271, 2000. ▶ 22, 23, 24, 25, 115, 156, 205, 209

[EK95] D.E. Edmunds and M. Krbec. Two limiting cases of Sobolev imbeddings. *Houston J. Math.*, 21:119–128, 1995. ▶ 4, 114, 156

[EKP00] D.E. Edmunds, R. Kerman, and L. Pick. Optimal Sobolev imbeddings involving rearrangement-invariant quasinorms. *J. Funct. Anal.*, 170:307–355, 2000. ▶ 4, 123

[EN98] D.E. Edmunds and Yu. Netrusov. Entropy numbers of embeddings of Sobolev spaces in Zygmund spaces. *Studia Math.*, 128(1):71–102, 1998. ▶ 206

[Enf73] P. Enflo. A counterexample to the approximation problem in Banach spaces. *Acta Math.*, 130:309–317, 1973. ▶ 198

[ET89] D.E. Edmunds and H. Triebel. Entropy numbers and approximation numbers in function spaces. *Proc. London Math. Soc.*, 58(3):137–152, 1989. ▶ 202

[ET92] D.E. Edmunds and H. Triebel. Entropy numbers and approximation numbers in function spaces II. *Proc. London Math. Soc.*, 64(3):153–169, 1992. ▶ 202

[ET95] D.E. Edmunds and H. Triebel. Logarithmic Sobolev spaces and their applications to spectral theory. *Proc. London Math. Soc.*, 71(3):333–371, 1995. ▶ 4

[ET96] D.E. Edmunds and H. Triebel. *Function Spaces, Entropy Numbers, Differential Operators*. Cambridge Univ. Press, Cambridge, 1996. ▶ 59, 110, 114, 198, 202, 206, 207, 209

[ET98] D.E. Edmunds and H. Triebel. Spectral theory for isotropic fractal drums. *C. R. Acad. Sci. Paris*, 326(11):1269–1274, 1998. ▶ 104

[ET99a] D.E. Edmunds and H. Triebel. Eigenfrequencies of isotropic fractal drums. *Oper. Theory Adv. Appl.*, 110:81–102, 1999. ▶ 104

[ET99b] D.E. Edmunds and H. Triebel. Sharp Sobolev embeddings and related Hardy inequalities: The critical case. *Math. Nachr.*, 207:79–92, 1999. ▶ 4, 135, 136, 137, 143

[FJ90] M. Frazier and B. Jawerth. A discrete transform and decomposition of distribution spaces. *J. Funct. Anal.*, 93(1):34–170, 1990. ▶ 104, 106, 120

[FL06] W. Farkas and H.-G. Leopold. Characterisation of function spaces of generalised smoothness. *Ann. Mat. Pura Appl.*, 185(1):1–62, 2006. ▶ 104

[FR04]   M. Frazier and S. Roudenko. Matrix-weighted Besov spaces and conditions of $\mathcal{A}_p$ type for $0 < p \leq 1$. *Indiana Univ. Math. J.*, 53(5):1225–1254, 2004. ▶ 127

[Fra86]   J. Franke. On the spaces $F^s_{p,q}$ of Triebel-Lizorkin type: Pointwise multipliers and spaces on domains. *Math. Nachr.*, 125:29–68, 1986. ▶ 110

[GCR85] J. García-Cuerva and J. L. Rubio de Francia. *Weighted Norm Inequalities and Related Topics*, volume 116 of *North-Holland Mathematics Studies*. North-Holland, Amsterdam, 1985. ▶ 61

[GKS87] Y. Gordon, H. König, and C. Schütt. Geometric and probabilistic estimates for entropy and approximation numbers of operators. *J. Approx. Theory*, 49:219–239, 1987. ▶ 199

[GNO04] A. Gogatishvili, J.S. Neves, and B. Opic. Optimality of embeddings of Bessel-potential-type spaces into Lorentz-Karamata spaces. *Proc. Roy. Soc. Edinburgh Sect. A*, 134(6):1127–1147, 2004. ▶ 124

[GNO05] A. Gogatishvili, J.S. Neves, and B. Opic. Optimality of embeddings of Bessel-potential-type spaces into generalized Hölder spaces. *Publ. Mat., Barc.*, 49(2):297–327, 2005. ▶ 157

[GO06]   P. Gurka and B. Opic. Sharp embeddings of Besov-type spaces. *J. Comput. Appl. Math.*, 2006. ▶ 124

[Gol79a] D. Goldberg. Local Hardy spaces. In: *Harmonic Analysis in Euclidean Spaces*, pages 245–248. Proc. Symp. Pure Math. 35, I, Amer. Math. Soc., Providence, R.I., 1979. ▶ 103

[Gol79b] D. Goldberg. A local version of real Hardy spaces. *Duke Math. J.*, 46:27–42, 1979. ▶ 103, 136

[Gol87a] M.L. Gol'dman. Imbedding theorems for anisotropic Nikol'skij-Besov spaces with moduli of continuity of general form. *Proc. Steklov Inst. Math.*, 170:95–116, 1987. translation from *Tr. Mat. Inst. Steklova*, 170:86-104, 1984. ▶ 123

[Gol87b] M.L. Gol'dman. On imbedding constructive and structural Lipschitz spaces in symmetric spaces. *Proc. Steklov Inst. Math.*, 173:93–118, 1987. translation from *Tr. Mat. Inst. Steklova*, 173:90-112, 1986. ▶ 123

[Gol87c] M.L. Gol'dman. On imbedding generalized Nikol'skij - Besov spaces in Lorentz spaces. *Proc. Steklov Inst. Math.*, 172(3):143–154, 1987. translation from *Tr. Mat. Inst. Steklova*, 172:128-139, 1985. ▶ 123

[Hal74] P.R. Halmos. *Measure Theory*, volume 18 of *GTM*. Springer, New York, 1974. Reprint of the ed. published by Van Nostrand, New York, 1950. ▶ 63

[Han79] K. Hansson. Imbedding theorems of Sobolev type in potential theory. *Math. Scand.*, 45:77–102, 1979. ▶ 4, 71, 88

[Har95] D. Haroske. Approximation numbers in some weighted function spaces. *J. Approx. Theory*, 83(1):104–136, 1995. ▶ 59

[Har97] D.D. Haroske. Embeddings of some weighted function spaces on $\mathbb{R}^n$; entropy and approximation numbers. A survey of some recent results. *An. Univ. Craiova, Ser. Mat. Inform.*, 24:1–44, 1997. ▶ 59

[Har98] D. Haroske. Some logarithmic function spaces, entropy numbers, applications to spectral theory. *Dissertationes Math.*, 373:59 pp., 1998. ▶ 17, 62, 209

[Har00a] D.D. Haroske. Logarithmic Sobolev spaces on $\mathbb{R}^n$; entropy numbers, and some application. *Forum Math.*, 12(3):257–313, 2000. ▶ 17, 209

[Har00b] D.D. Haroske. On more general Lipschitz spaces. *Z. Anal. Anwendungen*, 19(3):781–799, 2000. ▶ 22, 23, 24, 25, 94, 115, 116, 155, 156

[Har01] D.D. Haroske. Envelopes in function spaces - a first approach. Jenaer Schriften zur Mathematik und Informatik Math/Inf/16/01, 72 pp., Universität Jena, Germany, 2001. ▶ ix, 3, 135, 190

[Har02] D.D. Haroske. Limiting embeddings, entropy numbers and envelopes in function spaces. Habilitationsschrift, Friedrich-Schiller-Universität Jena, Germany, 2002. ▶ ix, 3, 113

[Harxx] D.D. Haroske. Growth envelope functions in Besov and Sobolev spaces. Local versus global results. *Math. Nachr.*, to appear. ▶ 171

[Hed72] L.I. Hedberg. On certain convolution inequalities. *Proc. Amer. Math. Soc.*, 36(2):505–510, 1972. ▶ 4

[Her68] C. Herz. Lipschitz spaces and Bernstein's theorem on absolutely convergent Fourier transforms. *J. Math. Mech.*, 18:283–324, 1968. ▶ 123

[HLP52] G. Hardy, J.E. Littlewood, and G. Pólya. *Inequalities*. Cambridge Univ. Press, Cambridge, 2nd edition, 1952. ▶ 108

[HM04] D.D. Haroske and S.D. Moura. Continuity envelopes of spaces of generalised smoothness, entropy and approximation numbers. *J. Approx. Theory*, 128(2):151–174, 2004. ▶ 116, 156, 181, 203

[HM06] D.D. Haroske and S.D. Moura. Continuity envelopes and sharp embeddings in spaces of generalised smoothness. Preprint, 2006. ▶ 157, 203

[HPxx] D.D. Haroske and I. Piotrowska. Atomic decompositions of function spaces with Muckenhoupt weights, and some relation to fractal analysis. *Math. Nachr.*, to appear. ▶ 127, 128, 176

[HS06] D.D. Haroske and L. Skrzypczak. Entropy and approximation numbers of embeddings of function spaces with Muckenhoupt weights. Preprint, 2006. ▶ 128, 187

[HT94a] D. Haroske and H. Triebel. Entropy numbers in weighted function spaces and eigenvalue distribution of some degenerate pseudo-differential operators I. *Math. Nachr.*, 167:131–156, 1994. ▶ 59, 124, 125

[HT94b] D. Haroske and H. Triebel. Entropy numbers in weighted function spaces and eigenvalue distribution of some degenerate pseudo-differential operators II. *Math. Nachr.*, 168:109–137, 1994. ▶ 59, 209

[HT05] D.D. Haroske and H. Triebel. Wavelet bases and entropy numbers in weighted function spaces. *Math. Nachr.*, 278(1-2):108–132, 2005. ▶ 59

[Jaw77] B. Jawerth. Some observations on Besov and Lizorkin-Triebel spaces. *Math. Scand.*, 40:94–104, 1977. ▶ 110

[JMP91] H. Johansson, L. Maligranda, and L.E. Persson. Inequalities for moduli of continuity and rearrangements. In: *Approximation Theory*, pages 413–423. Proc. Conf., Kecskemet/Hung. 1990, Colloq. Math. Soc. Janos Bolyai 58, North-Holland, Amsterdam, 1991. ▶ 195

[Kal82] G.A. Kalyabin. Criteria for the multiplicativity of spaces of Besov-Lizorkin- Triebel type and their imbedding into *C*. *Math. Notes*, 30:750–755, 1982. translation from *Mat. Zametki*, 30:517–526, 1981. ▶ 117

[KJF77] A. Kufner, O. John, and S. Fučík. *Function Spaces.* Academia, Publishing House of the Czechoslovak Academy of Sciences, Prague, 1977. ▶ 22

[KL87] G.A. Kalyabin and P.I. Lizorkin. Spaces of functions of generalized smoothness. *Math. Nachr.*, 133:7–32, 1987. ▶ 104, 123

[KLSS06a] Th. Kühn, H.-G. Leopold, W. Sickel, and L. Skrzypczak. Entropy numbers of embeddings of weighted Besov spaces. *Constr. Approx.*, 23:61–77, 2006. ▶ 59

[KLSS06b] Th. Kühn, H.-G. Leopold, W. Sickel, and L. Skrzypczak. Entropy numbers of embeddings of weighted Besov spaces II. *Proc. Edinburgh Math. Soc. (2)*, 49:331–359, 2006. ▶ 59, 128, 187

[KLSSxx] Th. Kühn, H.-G. Leopold, W. Sickel, and L. Skrzypczak. Entropy numbers of embeddings of weighted Besov spaces III. Weights of logarithmic type. *Math. Z.*, to appear. ▶ 59

[Kol89] V.I. Kolyada. Rearrangements of functions and embedding theorems. *Usp. Mat. Nauk*, 44(5(269)):61–95, 1989. Russian; English translation in: *Russ. Math. Surv.*, 44(5):73–117, 1989. ▶ 22, 190

[Kol98] V.I. Kolyada. Rearrangements of functions and embedding of anisotropic spaces of Sobolev type. *East J. Approx.*, 4(2):111–199, 1998. ▶ 123

[Kön86] H. König. *Eigenvalue Distribution of Compact Operators*. Birkhäuser, Basel, 1986. ▶ 198, 199, 208

[KPxx] R. Kerman and L. Pick. Optimal Sobolev imbeddings. *Forum Math.*, to appear. ▶ 71

[KS01] M. Krbec and H.-J. Schmeisser. Imbeddings of Brézis-Wainger type. The case of missing derivatives. *Proc. R. Soc. Edinb., Sect. A, Math.*, 131(3):667–700, 2001. ▶ 24, 156

[KS05] M. Krbec and H.-J. Schmeisser. Refined limiting imbeddings for Sobolev spaces of vector-valued functions. *J. Funct. Anal.*, 227:372–388, 2005. ▶ 142

[KS06] M. Krbec and H.J. Schmeisser. Critical imbeddings with multivariate rearrangements. Technical Report Math/Inf/04/06, p. 1-37, Universität Jena, Germany, 2006. ▶ 142

[Lan93] S. Lang. *Real and Functional Analysis*. Springer, New York, 1993. ▶ 63

[Leo98] H.-G. Leopold. Limiting embeddings and entropy numbers. Forschungsergebnisse Math/Inf/98/05, Universität Jena, Germany, 1998. ▶ 104, 184

[Leo00] H.-G. Leopold. Embeddings and entropy numbers in Besov spaces of generalized smoothness. In: H. Hudzik and L. Skrzypczak, editors, *Function Spaces*, pages 323–336. The Fifth Conference, Lecture Notes in Pure and Applied Math. 213, Marcel Dekker, 2000. ▶ 104, 203

[Lio98] P.-L. Lions. *Mathematical Topics in Fluid Mechanics. Vol. 2: Compressible Models*. Oxford Lecture Series in Mathematics and Its Applications, 10. Clarendon Press, Oxford, 1998. ▶ 25

[Liz86] P.I. Lizorkin. Spaces of generalized smoothness. *Mir, Moscow*, pages 381–415, 1986. Appendix to Russian ed. of [Tri83]; Russian. ▶ 104, 123

[Mar87] J. Marschall. Some remarks on Triebel spaces. *Studia Math.*, 87:79–92, 1987. ▶ 145

[Mar95] J. Marschall. On the boundedness and compactness of nonregular pseudo-differential operators. *Math. Nachr.*, 175:231–262, 1995. ▶ 145

[Maz72] V.G. Maz'ya. Certain integral inequalities for functions of several variables. In: *Problems of Mathematical Analysis, No. 3: Integral and Differential Operators, Differential Equations*, pages 33–68. Izdat. Leningrad. Univ., Leningrad, 1972. Russian; English translation in: *J. Soviet Math.*, 1:205–234, 1973. ▶ 4, 57, 71

[Maz85] V.G. Maz'ya. *Sobolev Spaces*. Springer Series in Soviet Mathematics. Springer, Berlin, 1985. Translated from the Russian by T.O. Shaposhnikova. ▶ 4, 26

[Maz05] V.G. Maz'ya. Conductor and capacitary inequalities for functions on topological spaces and their applications to Sobolev type imbeddings. *J. Funct. Anal.*, 224(2):408–430, 2005. ▶ 4

[Mil94] M. Milman. *Extrapolation and Optimal Decomposition*, volume 1580 of *LNM*. Springer, 1994. ▶ 23

[MNP06] S.D. Moura, J.S. Neves, and M. Piotrowski. Growth envelopes of anisotropic function spaces. Technical Report 06-18, Universidade de Coimbra, Portugal, 2006. ▶ 123, 142

[Mos71] J. Moser. A sharp form of an inequality by N. Trudinger. *Indiana Univ. Math. J.*, 20:1077–1092, 1970/71. ▶ 4, 57

[Mou01] S.D. Moura. Function spaces of generalised smoothness. *Dissertationes Math.*, 398:88 pp., 2001. ▶ 104

[Mou02] S.D. Moura. *Function spaces of generalised smoothness, entropy numbers, applications*. PhD thesis, Universidade de Coimbra, Portugal, 2002. ▶ 104, 116, 183

[Muc72] B. Muckenhoupt. Hardy's inequality with weights. *Studia Math.*, 44:31–38, 1972. ▶ 61

[Muc74] B. Muckenhoupt. The equivalence of two conditions for weight functions. *Studia Math.*, 49:101–106, 1973/74. ▶ 61

[Net87] Yu. Netrusov. Theorems of imbedding Besov spaces into ideal spaces. *Zap. Nauchn. Semin. Leningr. Otd. Mat. Inst. Steklova*, 159:69–82, 1987. Russian; English translation in: *J. Soviet Math.*, 47:2871–2881, 1989. ▶ 123

[Net89] Yu. Netrusov. Sets of singularities of functions in spaces of Besov and Lizorkin-Triebel type. *Trudy Mat. Inst. Steklov*, 187:162–177, 1989. Russian, English translation in: *Proc. Steklov Inst. Math.*, 187:185–203, 1990. ▶ 123

[Nev01a] J.S. Neves. Extrapolation results on general Besov-Hölder-Lipschitz spaces. *Math. Nachr.*, 230:117–141, 2001. ▶ 115, 155, 156

[Nev01b] J.S. Neves. *Fractional Sobolev-type spaces and embeddings.* PhD thesis, University of Sussex at Brighton, U.K., 2001. ▶ 83

[Pee66] J. Peetre. Espaces d'interpolation et théorème de Soboleff. *Ann. Inst. Fourier*, 16:279–317, 1966. ▶ 4, 70, 123, 143

[Pie80] A. Pietsch. *Operator Ideals.* North-Holland, Amsterdam, 1980. ▶ 198, 199

[Pie87] A. Pietsch. *Eigenvalues and s-Numbers.* Akad. Verlagsgesellschaft Geest & Portig, Leipzig, 1987. ▶ 198, 208

[Poh65] S. Pohožaev. On eigenfunctions of the equation $\Delta u + \lambda f(u) = 0$. *Dokl. Akad. Nauk SSR*, 165:36–39, 1965. Russian. ▶ 4, 57

[Pus01] E. Pustylnik. On some properties of generalized Marcinkiewicz spaces. *Studia Math.*, 144(3):227–243, 2001. ▶ 8

[Rou04] S. Roudenko. Duality of matrix-weighted Besov spaces. *Studia Math.*, 160(2):129–156, 2004. ▶ 127

[Ryc01] V.S. Rychkov. Littlewood-Paley theory and function spaces with A_p^{loc} weights. *Math. Nachr.*, 224:145–180, 2001. ▶ 61, 127

[Sch86] M. Schechter. *Spectra of Partial Differential Operators.* North-Holland, Amsterdam, 2nd edition, 1986. ▶ 208

[Skr05] L. Skrzypczak. On approximation numbers of Sobolev embeddings of weighted function spaces. *J. Approx. Theory*, 136:91–107, 2005. ▶ 59, 202

[Sob38] S.L. Sobolev. Sur un théorème d'analyse fonctionnelle. *Mat. Sb.*, N. Ser. 4 (46):471–497, 1938. Russian. French summary. English translation in: *Am. Math. Soc., Transl., II. Ser.*, 34:39–68, 1963. ▶ ix, 3, 35

[ST87] H.-J. Schmeißer and H. Triebel. *Topics in Fourier Analysis and Function Spaces.* Wiley, Chichester, 1987. ▶ 59

[ST95] W. Sickel and H. Triebel. Hölder inequalities and sharp embeddings in function spaces of $B_{p,q}^s$ and $F_{p,q}^s$ type. *Z. Anal. Anwendungen*, 14:105–140, 1995. ▶ 109, 110, 111

[Ste70] E.M. Stein. *Singular Integrals and Differential Properties of Functions.* Princeton University Press, Princeton, 1970. ▶ 190

[Ste81] E.M. Stein. Editor's note: The differentiability of functions on $\mathbb{R}^n$. *Ann. Math.*, II. Ser. 113:383–385, 1981. ▶ 190

[Ste93] E.M. Stein. *Harmonic Analysis.* Princeton University Press, Princeton, 1993. ▶ 61

[Str67] R.S. Strichartz. Multipliers on fractional Sobolev spaces. *J. Math. Mech.*, 16:1031–1060, 1967. ▶ 123

[Str72] R.S. Strichartz. A note on Trudingers extension of Sobolev's inequalities. *Indiana Univ. Math. J.*, 21:841–842, 1972. ▶ 4, 57

[Tal94] G. Talenti. Inequalities in rearrangement invariant function spaces. In: M. Krbec, A. Kufner, B. Opic, and J. Rákosník, editors, *Nonlinear Analysis, Function Spaces and Applications, Vol. 5*, pages 177–230. Spring School held in Prague, May, 1994, Prometheus Publishing House, 1994. ▶ 70

[Tar72] L. Tartar. Interpolation non linéaire et Regularité. *J. Funct. Anal.*, 9:469–489, 1972. ▶ 137

[Tar98] L. Tartar. Imbedding theorems of Sobolev spaces into Lorentz spaces. *Boll. Unione Mat. Ital., Sez. B, Artic. Ric. Mat. (8)*, 1(3):479–500, 1998. ▶ 123

[Tri78a] H. Triebel. *Interpolation Theory, Function Spaces, Differential Operators.* North-Holland, Amsterdam, 1978. ▶ 5, 61, 70, 120, 137, 206

[Tri78b] H. Triebel. *Spaces of Besov-Hardy-Sobolev type.* Teubner, Leipzig, 1978. ▶ 106

[Tri83] H. Triebel. *Theory of Function Spaces.* Birkhäuser, Basel, 1983. ▶ 5, 102, 103, 106, 107, 109, 115, 133, 136, 146, 206

[Tri92] H. Triebel. *Theory of Function Spaces II.* Birkhäuser, Basel, 1992. ▶ 5, 102, 125, 126, 173, 175

[Tri93] H. Triebel. Approximation numbers and entropy numbers of embeddings of fractional Besov-Sobolev spaces in Orlicz spaces. *Proc. London Math. Soc.*, 66(3):589–618, 1993. ▶ 4, 114, 143

[Tri94] H. Triebel. Relations between approximation numbers and entropy numbers. *J. Approx. Theory*, 78:112–116, 1994. ▶ 199

[Tri97] H. Triebel. *Fractals and Spectra.* Birkhäuser, Basel, 1997. ▶ 5, 102, 104, 105, 106, 207, 209

[Tri99] H. Triebel. Sharp Sobolev embeddings and related Hardy inequalities: The sub-critical case. *Math. Nachr.*, 208:167–178, 1999. ▶ 5, 121, 122, 123, 131, 132

[Tri01] H. Triebel. *The Structure of Functions*. Birkhäuser, Basel, 2001. ▶ ix, 5, 6, 44, 63, 64, 79, 83, 102, 112, 120, 122, 123, 135, 145, 153, 166, 170, 189, 192, 196

[Tri06] H. Triebel. *Theory of Function Spaces III*. Birkhäuser, Basel, 2006. ▶ 5, 102

[Tru67] N. S. Trudinger. On embeddings into Orlicz spaces and some applications. *J. Math. Mech.*, 17:473–483, 1967. ▶ 4, 57

[Vis98] M. Vishik. Hydrodynamics in Besov spaces. *Arch. Rational Mech. Anal.*, 145:197–214, 1998. ▶ 25

[Yud61] V.I. Yudovich. Some estimates connected with integral operators and with solutions of elliptic equations. *Soviet Math. Doklady*, 2:746–749, 1961. ▶ 57

[Zem80] J. Zemánek. The essential spectral radius and the Riesz part of the spectrum. In: *Functions, Series, Operators*. Proc. Int. Conf. Budapest 1980. *Colloq. Math. Soc. János Bolyai*, 35:1275–1289, North-Holland, Amsterdam, 1983. ▶ 207

[Zie89] W.P. Ziemer. *Weakly Differentiable Functions. Sobolev Spaces and Functions of Bounded Variation*, volume 120 of *GTM*. Springer, Berlin, 1989. ▶ 4, 26, 87, 123

[Zyg45] A. Zygmund. Smooth functions. *Duke Math. J.*, 12:47–76, 1945. ▶ 156

[Zyg77] A. Zygmund. *Trigonometric Series*. Cambridge Univ. Press, Cambridge, 2nd edition, 1977. ▶ 156

Symbols

Index

List of Figures